Brassey's
Battlefield Weapons Systems & Technology, Volume VII

Surveillance and Target Acquisition Systems

Brassey's
Battlefield Weapons Systems and Technology Series

General Editor: Colonel R G Lee OBE, Former Military Director of Studies at the Royal Military College of Science, Shrivenham, UK

This new series of course manuals is written by senior lecturing staff at RMCS, Shrivenham, one of the world's foremost institutions for military science and its application. It provides a clear and concise survey of the complex systems spectrum of modern ground warfare for officers-in-training and volunteer reserves throughout the English-speaking world.

Introduction to Battlefield Weapons Systems and Technology—R G Lee OBE

Volume I	Vehicles and Bridging—I F B Tytler *et al*
Volume II	Guns, Mortars and Rockets—J W Ryan
Volume III	Ammunition (Including Grenades and Mines)—K J W Goad and D H J Halsey
Volume IV	Nuclear, Biological and Chemical Warfare—L W McNaught and K P Clark
Volume V	Small Arms and Cannons—C J Marchant Smith and P R Haslam
Volume VI	Command, Control and Communications—A M Willcox, M G Slade and P A Ramsdale
Volume VII	Surveillance and Target Acquisition— A. L. Rodgers *et al.*
Volume VIII	Guided Weapons (Including Light, Unguided Anti-Tank Weapons) — R G Lee *et al.*
Volume IX	Military Data Processing and Microcomputers—J W D Ward and G N Turner
Volume X	Military Ballistics—A Basic Manual—C L Farrar and D W Leeming
Volume XI	Military Helicopters—P G Harrison

For full details of these titles in the series, please contact your local Brassey's/Pergamon office

Other Titles of Interest from Brassey's Defence Publishers

HEMSLEY	Soviet Troop Control. The Role of Automation in Military Command
SIMPKIN	Human Factors in Mechanized Warfare
SORRELS	US Cruise Missile Programs: Development, Deployment and Implications for Arms Control

Surveillance and Target Acquisition Systems

A. L. RODGERS
I. B. R. FOWLER
T. K. GARLAND-COLLINS
J. A. GOULD
D. A. JAMES
W. ROPER

Royal Military College of Science, Shrivenham, UK

BRASSEY'S DEFENCE PUBLISHERS
a member of the Pergamon Group

OXFORD · NEW YORK · TORONTO
SYDNEY · PARIS · FRANKFURT

U.K.	BRASSEY'S PUBLISHERS LTD a member of the Pergamon Group Headington Hill Hall, Oxford OX3 0BW, England
U.S.A.	Pergamon Press Inc., Maxwell House, Fairview Park, Elmsford, New York 10523, U.S.A.
CANADA	Pergamon Press Canada Ltd., Suite 104, 150 Consumers Road, Willowdale, Ontario M2J 1P9, Canada
AUSTRALIA	Pergamon Press (Aust.) Pty. Ltd., P.O. Box 544, Potts Point, N.S.W. 2011, Australia
FRANCE	Pergamon Press SARL, 24 rue des Ecoles, 75240 Paris, Cedex 05, France
FEDERAL REPUBLIC OF GERMANY	Pergamon Press GmbH, Hammerweg 6, D-6242 Kronberg-Taunus, Federal Republic of Germany

First edition 1983

Library of Congress Cataloging in Publication Data
Main entry under title:
Surveillance & target acquisition systems.
(Battlefield weapons systems & technology; v. 7)
Includes index.
1. Military surveillance—Equipment and supplies—Handbooks, manuals, etc. 2. Target acquisition—Equipment and supplies—Handbooks, manuals, etc.
I. Rodgers, A. L. II. Title: Surveillance and target acquisition systems. III. Series.
UG475.S97 1983 623.7 83-8824

British Library Cataloguing in Publication Data
Surveillance and target acquisition systems.
—(Brassey's battlefield weapons systems and technology series; v.7)
1. Armies—Equipment 2. Electrooptical devices
I. Rodgers, A.L.
623'.042 UF849

ISBN 0-08-028334-9 (Hardcover)
ISBN 0-08-028335-7 (Flexicover)

In order to make this volume available as economically and as rapidly as possible the authors' typescripts have been reproduced in their original forms. This method unfortunately has its typographical limitations but it is hoped that they in no way distract the reader.

The views expressed in the book are those of the authors and not necessarily those of the Ministry of Defence of the United Kingdom.

Printed in Great Britain by A. Wheaton & Co. Ltd., Exeter

Preface

The Series

This series of books is written for those who wish to improve their knowledge of military weapons and equipment. It is equally relevant to professional soldiers, those involved in developing or producing military weapons or indeed anyone interested in the art of modern warfare.

All the texts are written in a way which assumes no mathematical knowledge and no more technical depth than would be gleaned from school days. It is intended that the books should be of particular interest to army officers who are studying for promotion examinations, furthering their knowledge at specialist arms schools or attending command and staff schools.

The authors of the books are all members of the staff of the Royal Military College of Science, Shrivenham, which is comprised of a unique blend of academic and military experts. They are not only leaders in the technology of their subjects, but are aware of what the military practitioner needs to know. It is difficult to imagine any group of persons more fitted to write about the application of technology to the battlefield.

Volume VII

The ability to carry out surveillance and acquire targets in poor visibility conditions and at night has taken a giant step forward in recent years. It has been helped by the advances made in microelectronics. This volume covers the scientific techniques used: it spans optics, image intensification, thermal imaging, radar and lasers. It also goes on to describe the battlefield use of these techniques.

Shrivenham, November 1982 — Geoffrey Lee

Acknowledgements

The authors wish to acknowledge the assistance of present and previous members of the military, academic and support staff, either directly or by making teaching precis available. In particular they wish to record their appreciation of Mr. D. K. Thomas for valued comment, to Mr. H.A.E. Summerfield for graphics and to Mrs. L. Rotherham for typing.

ALR
WR
IBRF
TKG-C
DAJ
JAG

Shrivenham
December 1982

Contents

List of Illustrations		ix
List of Tables		xv
Chapter 1	Introduction	1
Chapter 2	Optics	17
Chapter 3	Image Intensification	51
Chapter 4	Thermal Imagers	69
Chapter 5	Lasers	91
Chapter 6	Radar	125
Chapter 7	Surveillance in Depth	157
Chapter 8	Counter Surveillance	173
Answers to Self Test Questions		187
Glossary		199
Index		213

List of Illustrations

Chapter 1

FIG 1.1	Location of targets	5
FIG 1.2	Intervisibility in NW Europe	8
FIG 1.3	Meteorological visibility in W Germany (daylight)	8
FIG 1.4	Surveillance by tethered platform and aerial vehicle	11
FIG 1.5	Typical remote sensor systems	11

Chapter 2

FIG 2.1	Cross section of the human eye	18
FIG 2.2	Dark adaptation	20
FIG 2.3	Dark adaptation as a function of contrast ratio	21
FIG 2.4	Wavelength response of rods and cones	22
FIG 2.5	Pupil reflex	23
FIG 2.6	Variation of acuity of the eye and pupil diameter	24
FIG 2.7	Variation of visual acuity with object contrast	25
FIG 2.8	Light contrast capability of the eye	27
FIG 2.9	Improvement of eye performance at low light levels	30
FIG 2.10	Main parameters of an optical system	32
FIG 2.11	Bar pattern analysis criteria	33
FIG 2.12	Optical Transfer Function for human eye	34
FIG 2.13	Small arms sight	36

FIG 2.14 Prismatic binocular 37

FIG 2.15 AFV optics 41

FIG 2.16 Illuminance and eye stimulus 45

Chapter 3

FIG 3.1 Spectral distribution characteristics of sunlight, moonlight, starlight and some blackbodies 53

FIG 3.2 Image intensifier (Diagram by kind permission of Mullards Limited) 54

FIG 3.3 Photocathode performance 55

FIG 3.4 Three stage (cascade) tube (Diagram by kind permission of Mullard Limited) 57

FIG 3.5 Channel electron multiplication 58

FIG 3.6 Channel image intensifiers 58

FIG 3.7 UK NOD A and IWS 59

FIG 3.8 Vidicon television camera tube 61

FIG 3.9 Hele-Tele in stabilised mount (Photo by courtesy of Marconi Avionics Limited) 63

FIG 3.10 Reflectivity of natural vegetation 65

Chapter 4

FIG 4.1 Blackbody radiation 70

FIG 4.2 Atmospheric windows 73

FIG 4.3 Energy level diagram for photoconduction 75

FIG 4.4 Response of infra-red detectors 77

FIG 4.5 Block schematic of thermal imager 79

FIG 4.6 Simplified single-detector, dual-axis scanner 80

FIG 4.7 Parallel scan processing 81

FIG 4.8 Parallel scanning mechanism 81

FIG 4.9 Serial scan processing 82

FIG 4.10 Pyro-electric vidicon 83

FIG 4.11 Class II thermal imager (Photo by courtesy of Marconi Avionics Limited) 84

FIG 4.12 Landrover viewed through a cooled thermal imager (Photo by courtesy of Marconi Avionics Limited) 85

Chapter 5

FIG 5.1 Spontaneous emission 92

FIG 5.2 Stimulated emission 93

FIG 5.3 Concept of cavity resonator 94

FIG 5.4 Pumping cycle for a 4-level laser 98

FIG 5.5 Helium Neon Laser 99

FIG 5.6 Principle of operation of the ruby (and neodymium) laser 101

FIG 5.7 Two types of Q-switch 102

FIG 5.8 Semi-conductor laser 104

FIG 5.9 Principle of pulsed laser range finder 108

FIG 5.10 CO_2 laser range finders (Ferranti Type 307 - Photo by courtesy of Ferranti) (Marconi Mark III - Photo by courtesy of Marconi Avionics Limited) 111

FIG 5.11 Target designator concept 112

FIG 5.12 LTMR and LRMTS (Photo by courtesy of Ferranti) 113

FIG 5.13 Doppler principle 116

FIG 5.14 Doppler system 117

Chapter 6

FIG 6.1 Pulse range measurement 126

FIG 6.2 CRO method of range measurement (Diagram is reproduced from AP3302 Part 3, Radar) 126

FIG 6.3 Digital range measurement 127

FIG 6.4 FMCW range measurement 128

FIG 6.5 Direction finding by narrow beam (Diagram is reproduced from AP3302 Part 3, Radar) 128

FIG 6.6 Producing a fan beam (Diagram is reproduced from AP3302 Part 3, Radar) 129

FIG 6.7 Air defence, vertical coverage diagram ((a) Diagram by courtesy of Plessey Radar) 130

FIG 6.8 Cosecant^2 antenna (Photo by courtesy of Marconi Radar) 131

FIG 6.9 PPI display (Diagram is reproduced from AP3302 Part 3, Radar) 131

FIG 6.10 Air defence surveillance radar - block diagram 132

FIG 6.11 Height finding radar ((a) and (c) Diagram is reproduced from AP3302 Part 3, Radar; (b) Photo by courtesy of Marconi Radar) 133

FIG 6.12 Multiple beams in elevation 134

FIG 6.13 Synthetic PPI display (Photo by courtesy of Plessey Radar) 134

FIG 6.14 Lightweight array antenna for small radars (Photo by courtesy of Marconi Avionics) 135

FIG 6.15	Mortar location radar (Photo by courtesy of EMI Electronics Ltd)	136
FIG 6.16	Atmospheric attenuation	138
FIG 6.17	Horizon	140
FIG 6.18	Plane earth reflection	140
FIG 6.19	Atmospheric ducting	141
FIG 6.20	The Doppler effect	141
FIG 6.21	Doppler radar	142
FIG 6.22	Pulse radar with Doppler MTI	142
FIG 6.23	PPI with MTI (Diagram is reproduced from AP3302 Part 3, Radar)	143
FIG 6.24	SLAR (Diagram is reproduced from AP3302 Part 3, Radar)	144
FIG 6.25	Nimrod in AEW configuration (Photo by courtesy of British Aerospace)	145
FIG 6.26	Radar tracking in angle (one plane)	146
FIG 6.27	Conical span (Diagram is reproduced from AP3302 Part 3, Radar)	146
FIG 6.28	Secondary radar (IFF)	147
FIG 6.29	ATC display (Diagram is reproduced from AP3302 Part 3, Radar)	148
FIG 6.30	Electronic beam steering	149
FIG 6.31	A phased array (Photo by courtesy of Raytheon)	149
FIG 6.32	Multiple beam formation	150

FIG 6. 33 Electronic beam steering in elevation (Photo by courtesy of Plessey Radar) 150

FIG 6. 34 Pulse compression 151

Chapter 7

FIG 7. 1 Surveillance sensors 158

FIG 7. 2 KIEBITZ (Photo by courtesy of Dornier Ltd) 160

FIG 7. 3 The SOTAS concept 161

FIG 7. 4 The MIDGE drone at launch (Photo by courtesy of Canadair Ltd) 163

FIG 7. 5 Battlefield tasks 165

FIG 7. 6 The Canadair CL 227 RPH (Photo by courtesy of Canadair Ltd) 166

FIG 7. 7 Principal elements of RPH and control station 167

FIG 7. 8 Schematic layout of a RGS system 170

Chapter 8

FIG 8. 1 Armoured vehicle firing smoke projectors (Photo by courtesy of ATDU, Bovington) 176

List of Tables

Chapter 1

TABLE 1 Areas of Interest and Influence Related to Weapon Systems 2

Chapter 3

TABLE 1 Illuminance Levels by Day and Night in Lux and Foot Candles 52

Chapter 4

TABLE 1 Critical Wavelengths (λ_c) below which Photo Conduction will occur for some Semi-conductors 74

Chapter 5

TABLE 1 Laser Beam Diameter in cm as a function of Range 94

TABLE 2 Aerosol Sizes 96

TABLE 3 Protection Standards 105

TABLE 4 Nominal Ocular Hazard Distances 106

TABLE 5 Two Field Type Laser Range Finders 109

Chapter 6

TABLE 1 Typical Values of σ 137

1.

Introduction

INTRODUCTION

Surveillance is a word with which everyone is familiar. Hardly a day goes by without a television programme in which the police or secret service keep a subject under surveillance. The subject really made the headlines during the Watergate affair when a 'bugging' or electronic surveillance attempt was discovered. As far as an Army is concerned, it requires early information about enemy dispositions and intentions so that forces can be positioned in the best way to react to the threat. In the context of General War, the reaction must take place as early as possible to enable the process of attrition to start at the maximum range. There is thus a need for a general surveillance capability progressing naturally into target acquisition and the engagement of those targets.

A dictionary definition of surveillance is "supervision, close observation". The definition which will be used throughout this book is "The continuous (all weather, day and night) systematic watch over the battlefield to provide timely information for combat intelligence". It is the continuous, all weather requirement, and the need for timely information, implying long range, which presents the challenge to technology.

This chapter will examine the problems encountered in providing an adequate system of surveillance and target acquisition and outline the techniques which can be used to overcome these problems.

DEFINITIONS

There are a number of agreed terms and definitions which are set out in the glossary. It is important that there is agreement on terminology so that performance specifications can be clearly and unequivocably understood.

The basic need of the commander, at whatever level, is for combat intelligence. He needs to know about the enemy, the weather and geographical features so that he can plan and conduct his operations. This fundamental need is there, for the infantry section commander as much as the corps commander. The significant difference between the requirements for these two very different individuals is how far into the enemy position their surveillance needs to penetrate. Obviously the corps commander will want to know what is going on well to the rear of the enemy position. The section commander needs to know what is going on a few hundred metres in front of him. Thus it is important to define a commander's Area of Interest and Area of Influence.

The commander's Area of Interest is perhaps the more difficult of the two areas to delimit. The area includes areas occupied by enemy forces which could jeopardise the accomplishment of the mission. This could mean almost anything but is usually taken to mean the reserve of the commander's opposite number. Thus a divisional commander will be interested in the deployment and intentions of the second echelon regiments of the Warsaw Pact division immediately to his front and this will extend up to some 150 km beyond the FEBA. Similarly the second echelon divisions of the opposing army will be in the Area of Interest of the corps commander, probably an area some 200 km deep.

Areas of Influence are much easier to define. Put in the simplest terms they are the areas over which a commander can bring fire to bear. For a section commander this will be a few hundred metres. For a corps commander, the maximum range of his SSM system will probably be in excess of 100 km. Table 1 shows Areas of Influence and Interest related to the general capabilities of current NATO weapons.

TABLE 1 Areas of Interest and Influence Related to Weapon Systems

Formation	Weapon	Range km	Area of Influence km	Area of Interest km
Battlegroup	MBT	2	6	20
	Mortar	6		
	LRATGW	4		
Brigade	105 mm SP	15	20	50
	155 mm SP	20		
Division	SSM	100	100	150
Corps	SSM	100	100	200

Table 1 Areas of Interest and Influence Related to Weapon Systems

The ranges quoted in Table 1 are for no particular weapon system. They are typical of the sort of ranges of equipments in service with most modern armies. They are fundamental to the consideration of surveillance and target acquisition since the surveillance system available to a commander should cover his area of interest otherwise he must fight at a disadvantage - like a short sighted boxer. Similarly the range of a target acquisition system must match the range of the weapon. It is pointless spending large sums of money on bringing long range weapons into service if that long range cannot be exploited because potential targets are out of range of the target acquisition system.

Having defined the basic requirements for surveillance and target acquisition it is necessary to consider in general terms how a target is acquired. There are 4 stages in target acquisition:

Detection
Recognition
Identification
Location

Detection

Detection is the discovery of a potential target. It is accomplished because of the contrast between the target and its background or some discontinuity with its surroundings. The observer's reaction is "There's something over there".

Recognition

Recognition is the determination of the class of the target. The observer's reaction is "It's a tank". A target can be recognised by its appearance or its behaviour. The distinguishing features which determine the appearance of the target are important in both recognition and identification. For example, tracks as opposed to wheels normally indicate that the object is an APC, gun or tank; the presence or absence of a turret and gun will help to decide exactly which class of vehicle it is. Help in recognition of a target may also be gained from its behaviour, such as the tactics which it adopts, its location or the direction of its movement. It may have to be observed for some time before a discernable pattern of behaviour emerges. Sometimes a significant delay between detection and recognition is inherent in both these methods. Trials have shown that the relationship between visual detection and recognition ranges is of the order of 6:1.

Identification

Identification is the stage in the acquisition process in which the true identity of the target is established. The observer's reaction is: "It's a T62". Although the characteristic features of such a tank may indicate its nationality or distinguish it from all other equipment in its class they are often minor and easily obscured - for instance the number of running wheels on a tank. Careful scrutiny of the target is required before a positive identification can be made.

The problems of identification can be eased by the use of various schemes. There may be some overt indication: the display of some readily recognisable feature such as a call sign, national distinguishing mark or air recognition sign could be examples of this.

On the other hand a covert method of identification is often used. An interrogation or challenge and reply sequence is a common method which may be audible using passwords; visual with a prearranged system of very lights or electronic which is a technique employed in many air defence systems. Immediately a target is detected by a radar it is challenged by a coded radar signal. Friendly aircraft are equipped with a transponder which decodes the challenge and replies. The code used in the reply indicates the identity of the aircraft. Such a system is called Identification - Friend or Foe (IFF).

Location

A target is located when its position is fixed with the accuracy needed for a successful engagement. The location of targets for a direct fire weapon is a relatively simple operation but with indirect fire weapons the procedure is more demanding. The differences are illustrated in Fig. 1.1.

In the case of direct fire systems, a direct fire weapon is aligned on the target, in azimuth and elevation, using the sighting system. Azimuth is provided accurately by the sight; the range to the target must be determined so that the correct elevation can be applied. There are a number of techniques available, most of which have been made obsolete by the introduction of the laser range finder. The most basic is visual range estimation in which range can be estimated visually, with or without the aid of a map. In stadiametric ranging a graticule in the sight allows ranging of targets of known height and width. Such a system is still in use on the T54 tank and on Leopard as a back up system. Ranging rounds can be used which involves firing the main armament. It is slow, reveals the weapon and unlikely to be used in a modern system. A development of this is to use a ranging machine gun. It is a system which has been widely used on British tanks and anti-tank guns. There are a number of different types of optical range finders. The stereoscopic type of range finder is still used but the simpler coincidence type is preferred in the M60 and AMX30 tanks. Radar is an important surveillance system in its own right. An important attribute of most radar systems is that the range to the target can be measured accurately. However, the introduction into service of the laser range finder in the last decade has overshadowed other developments in this field. It is easy to use and is highly accurate.

In indirect fire systems the problem of location is aggravated by the absence of a line of sight between the target and the weapon. The observer can locate the target, relative to his own position, using the same sort of techniques as for direct fire weapons. To transfer this information to the weapon it is necessary to know its position and that of the observer to the required degree of accuracy.

Conventional map reading is not accurate enough for field artillery; position determining equipment is needed.

In certain guided weapon systems after a target has been acquired, the positions of both missile and target are monitored continuously. The flight path of the missile is then corrected to achieve impact. There are several ways of doing this: optical, infra-red or radar can be used. In some systems there is a combination of methods.

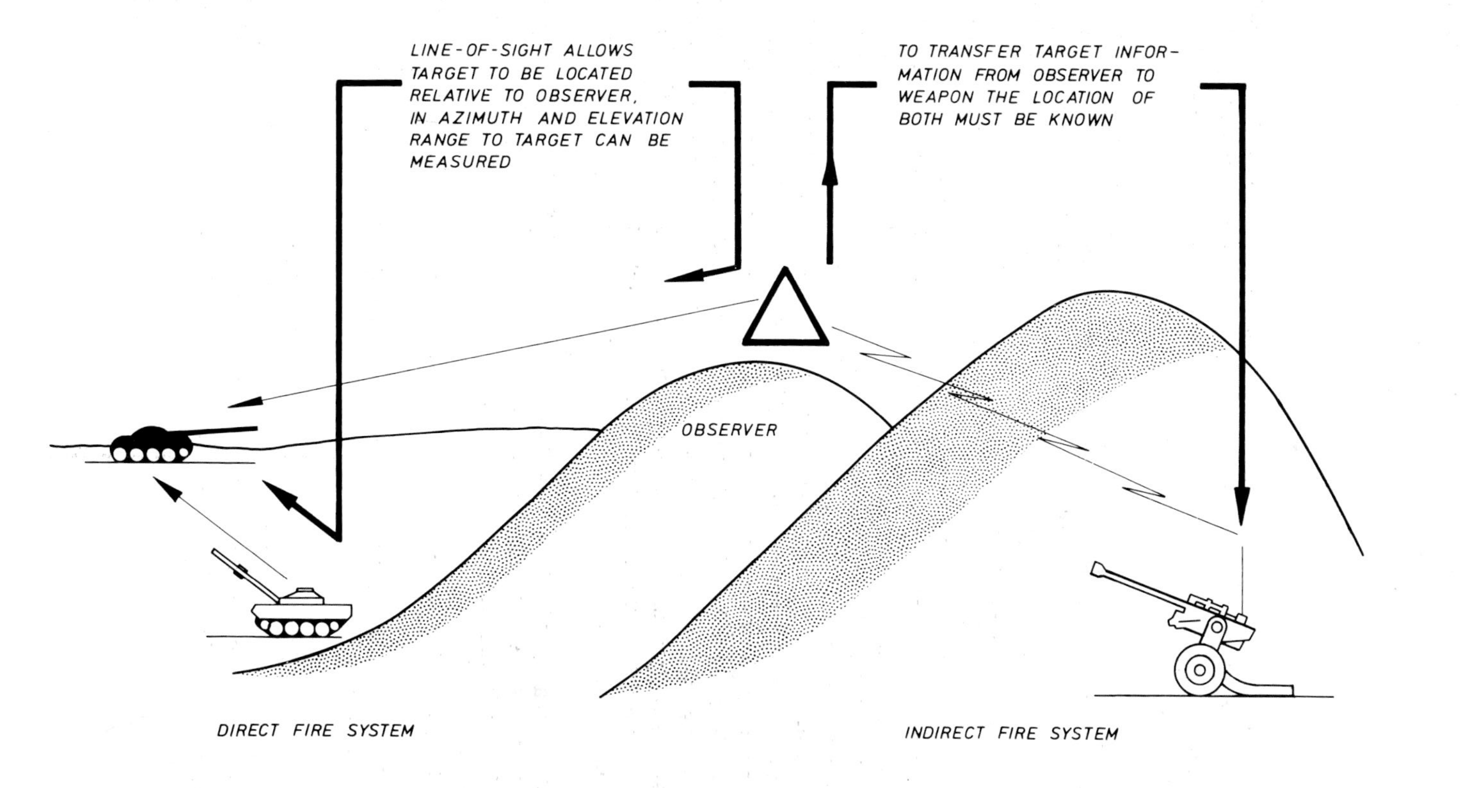

Fig. 1.1 Location of targets

TECHNIQUES USED IN SURVEILLANCE AND TARGET ACQUISITION

Human Senses

A target is likely to reveal its presence because of characteristics which are inherent either in its design or its tactical employment. In the simplest case such characteristics can be recognised by the five human senses. As an example a tank is most likely to be detected and recognised by sight and hearing. It has a shape which can be seen and with training, identified. The power plant, transmission and running gear of the tank make a noise which will disclose its presence and give some idea of the type of vehicle. It also emits exhaust fumes which can be smelled, but this is normally a short range effect as are touch and taste which are unlikely to be used in the detection of enemy armour.

Target Emissions

For centuries man has endeavoured to extend the range of his senses by use of telescopes, binoculars and other similar devices. It therefore follows that it may be possible, by using suitable instruments, to detect and recognise a tank by the vibrations which it sets up in the ground, the way in which it disturbs the earth's magnetic field, the natural radiation which it emits (in the case of the tank this will be infra-red radiation), or emissions from its ancilliary equipment such as radios and searchlights. The most widely used methods of detecting targets at present are those which make use of electro magnetic radiation. These methods can be sub-divided according to the source of the energy involved.

A wide range of frequencies may be emitted by a target. Some emissions are unavoidable and come about because any object at a temperature above absolute zero radiates energy. It is known as black body radiation. For targets of military interest most of this radiation will be infra-red although some energy is also emitted in the microwave part of the spectrum and could be of use in detecting aircraft. Infra-red radiation is used in such systems as thermal imaging and infra-red linescan. Other electro magnetic radiation is emitted by the target due to its operational employment. This includes light and infra-red emissions as well as radio and radar transmissions.

Reflected Energy

Energy radiated by an external source may be reflected by the target. Most of the energy radiated by the sun occurs in the visible portion of the spectrum as light. This can be detected after reflection by the target by all optical systems. It is the way in which we normally see any object. There is also considerable radiation in the near infra-red portion of the spectrum and this is utilised in image intensifiers and low light television systems. The performance of optical and near infra-red systems can be enhanced by the use of artificially generated radiation. Both light and near infra-red energy can be produced by searchlights, flares and the like in order to illuminate targets at night. In a similar way a target may be illuminated by a radar with electro magnetic radiation at radar frequencies.

Other Techniques

Although in land systems most effort is at present concentrated on the detection of targets by electro magnetic effects, there is increasing interest in making use of energy propagated in other ways such as elastic waves. The prime example of this is making use of the seismic waves set up in the earth by a moving object or by an explosion. Acoustic waves are also important and are used in sound ranging systems for locating guns and in sniper locating. At sea, for example in the search for submarines, this type of technique assumes a much greater importance than on land.

Many surveillance and target acquisition problems cannot be solved by the methods so far outlined. For example, the discovery of hidden arms, ammunition or explosives in an internal security situation. In this type of environment other senses and sensors must be brought into use. Examples of this are tracker dogs using their sense of smell and sniffer devices for the detection of hidden personnel or objects.

ACTIVE AND PASSIVE SYSTEMS

An active surveillance system radiates energy at the target in order to illuminate it. The use of searchlights, lasers and radars are examples and it is clear that these and other active techniques can themselves be detected and thus reveal the position of the observer and lay themselves open to countermeasures.

Passive systems do not rely on the radiation generated by the observer. They therefore tend to consume less power, are less likely to give away the position of the observer but are still liable to countermeasures. For example, a passive homing missile may be decoyed by the use of infra-red flares dropped by the target aircraft. Nevertheless great emphasis is currently placed on the replacement of active systems by passive systems wherever possible.

THE LIMITATIONS OF ACQUISITION RANGES

Governing Factors

The range at which a target is detected and subsequently recognised is governed by terrain screening or intervisibility, the characteristics of the target and meteorological conditions.

Terrain Screening

Most surveillance devices require a line of sight to the target; natural and artificial obstacles, such as hills and villages will shield a target from observation.

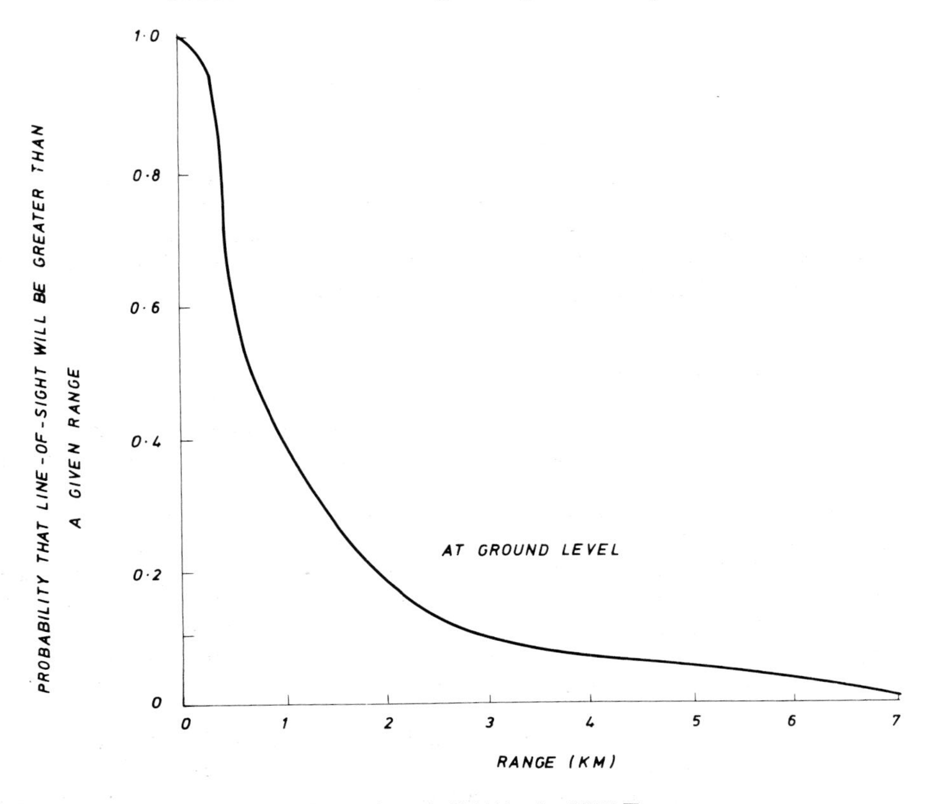

Fig. 1.2 Intervisibility in NW Europe

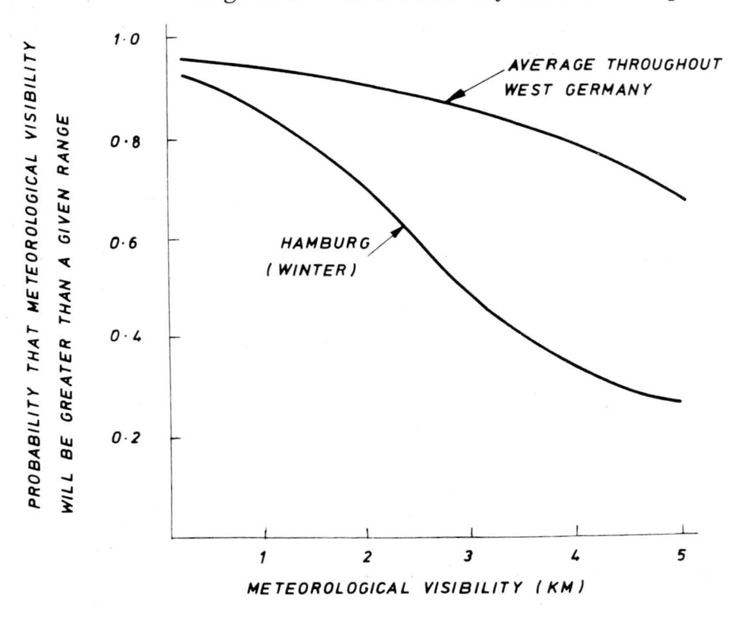

Fig. 1.3 Meteorological visibility in W Germany (daylight)

In certain types of terrain, such as the desert, terrain screening may not be significant but in a typical NW European environment it is a major constraint on acquisition ranges. The conditions of intervisibility in NW Europe are shown in Fig. 1.2; it can be seen that, at ground level and in the most favourable meteorological conditions, there is a less than 10% chance of being able to see more than 3 km and there is a less than 5% chance of being able to see more than 5 km. There is therefore little justification for a ground based surveillance system with a range greater than 5 km. An extended line of sight can often be obtained by siting the observer on high ground. But the ground overlooked may not be of tactical significance and inevitably there will be areas of dead ground.

Characteristics of the Target

The range at which a target can be acquired is governed by the contrast between the target and its surroundings and the physical characteristics of the target.

Initially the presence of a target is disclosed by the difference between its features and those of its background, this is its contrast. The higher this contrast the more easily is the target acquired. In consequence, a target which presents a high contrast can be seen at a greater range than one which blends well with its surroundings. The contrast must be evident at the wavelength at which the surveillance device operates. For example, a device which is conspicuous when viewed by a thermal device may be well concealed from visual observation.

The physical characteristics which most affect the acquisition range are the energy it radiates or reflects and its size. In general terms the higher the level of energy emitted or reflected by the target the greater the range at which it will be seen. Thus the size of the target and the nature of its surface which affects its ability to radiate or reflect energy (emissivity and reflectivity) are significant factors in determining acquisition ranges.

Meteorological Conditions

Poor light conditions are an obvious problem. Visual surveillance is particularly ineffective in darkness. Simple optical instruments such as weapon sights and binoculars will provide some enhancements when conditions are as bright as clear moonlight. But their performance in these circumstances does not compare with their capability in daylight: acquisition ranges are reduced by a factor of about ten. The performance of passive devices which make use of night sky radiation is also governed by the level of light. Such devices cannot provide day and night surveillance unless a source of illumination is available for use on dark, overcast nights.

Adverse weather conditions often degrade surveillance. The performance of all surveillance devices is affected to some extent by adverse weather. Fog, smoke and rain absorb the radiation from the target and degrade the contrast which the target creates. The susceptibility to atmospheric interference of a surveillance system depends on the wavelength at which the system operates. The longer the wavelength, the less is the system affected. The graph in Fig. 1.3 illustrates the extent to which the performance of visual systems is affected. Although the overall effect is not serious, atmospheric interference can be severe in certain areas, for example over low-lying ground, particularly in winter. Radar which operates at wavelengths very much longer than visible light is much less sensitive to atmospheric conditions.

A system which can work throughout the 24 hours is required. During daylight in clear weather an optical system is to be preferred because its resolution is better than any other form of surveillance device and hence it will achieve recognition and identification at greater ranges than other systems. It is also easy for the operator to interpret the image. But the optical system will not perform in the dark, so some other device such as an image intensifier, low light television or thermal imager is needed, depending on the circumstances. In their turn these are degraded by bad weather so a radar is needed to ensure an all weather capability. But the resolution of the radar is such that it has only a limited recognition capability. It can tell the difference between a tracked vehicle, a wheeled vehicle and a man, but not the difference between different types of tracked vehicles. Thus to provide an all weather, day or night capability no one technique is sufficient and a balanced mix of techniques is required.

BATTLEFIELD SURVEILLANCE IN DEPTH

For target acquisition purposes an adequate system of surveillance is required out to the depth of the area of influence, but general surveillance is also required out to the limits of the area of interest. At divisional level these might be 100 km and 150 km respectively. Thus it is clear that the effects of terrain screening must be overcome. Simply put there are two ways of doing this; either elevate the sensor or move it closer to the target. The line of sight can be extended by elevating the surveillance device above ground level; intervisibility improves as the height increases. Vertical observation provides the most complete coverage as at low elevations reverse slopes may be concealed from view, see Fig. 1.4. Areas which are screened from observation can be monitored by troops or surveillance devices positioned in dead ground: this method is often referred to as Remote Ground-Based Surveillance.

Aerial Surveillance

The types of aerial vehicles likely to be available for surveillance are mentioned below. Satellites are not included since strictly speaking they are not aerial vehicles. However, satellites are clearly of the utmost importance in attaining strategic information. In the foreseeable future they are unlikely to be a regular reliable source of battlefield intelligence in all weather conditions, so manned and unmanned aerial vehicles will still be important.

High performance, manned aircraft are capable of deep penetration and low level reconnaissance. At present the information acquired by the sensors is recorded on film for subsequent exploitation. The information is not transmitted in-flight although this is technically possible. Thus the time from requesting a mission to receiving the intelligence - the response time - is a matter of hours.

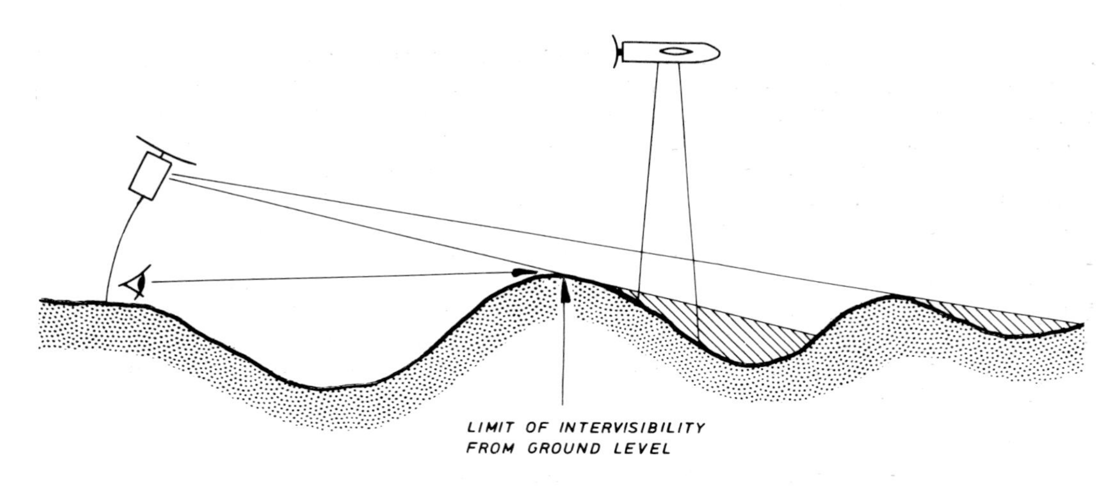

Fig. 1.4 Surveillance by tethered platform and aerial vehicle

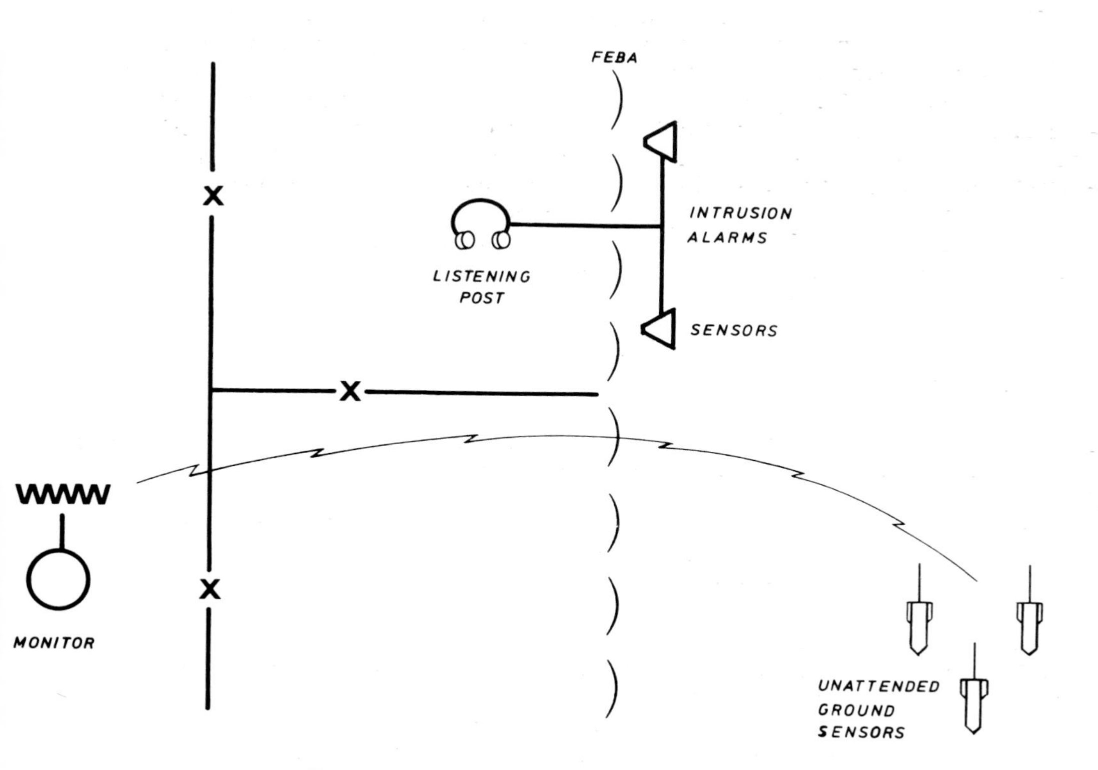

Fig. 1.5 Typical remote sensor systems

Modern helicopters have greatly improved day, night and all weather performance. Optical aids such as stabilised sights and binoculars are available for daytime use while for night use, image intensifier goggles and low light television are available and thermal imagers could well be used in future. Nevertheless, helicopters are vulnerable and are unlikely to be used forward of the FLOT. Thus their ability to see deep into the enemy position will be limited.

Tethered platforms or masts are a means of elevating surveillance devices above ground level. Several advanced techniques have been tried but none has so far proved to be successful; instability in wind is a severe problem.

A drone is an unmanned aerial vehicle which flies a pre-set flight path. Like the manned reconnaissance aircraft it records the information which it acquires on film for subsequent analysis. It is possible for the information to be transmitted in flight. The drone is relatively invulnerable and has a fairly good response time. Its major drawback is inflexibility. There is no way of altering the flight path once the vehicle has been launched. Modern drones are capable of penetrating about 50 km beyond the FLOT.

An RPV is an unmanned aerial vehicle the flight path of which can be controlled by a ground or airborne pilot. It is far more flexible than a drone because the surveillance sensor output can be monitored and the vehicle commanded to fly whereever the controller wishes and carry out whatever task he demands. It is a suitable vehicle to fly over a target to correct artillery fire by electro-optical means. Terrain screening is a limitation on the distance that the pilot can control the RPV because of the need to maintain a communication link with it. An RPV system will be significantly more expensive than a drone system.

Remote Ground-Based Surveillance

Patrols, OPs and stay behind parties have always been used to acquire information from behind the enemy's front line. They will continue to play their part in the surveillance system. Their employment is a tactical matter outside the scope of this book. They will almost certainly use surveillance equipment to help them to do their job. However, the last 20 years or so has seen the development of remote ground sensors (RGS). These unmanned systems fall into two categories; Intrusion Alarms to detect enemy penetration into friendly areas and Unattended Ground Sensors to monitor enemy activity in depth.

An unmanned sensor system is shown in diagramatic form in Fig. 1.5. The elements of such a system are a sensor or sensors, a communication link and a monitoring system.

A variety of techniques can be used, for example, seismic, acoustic, radar or infra-red. In the simplest system each sensor is coupled directly to a communication link by which the sensor is monitored continuously. In more sophisticated systems the sensor analyses the information and sends processed information to the monitor.

The sensor or sensors can be connected to the monitor by a communication link using either line or radio. Over long transmission paths it may be necessary to provide relays.

The information provided by the sensors can be displayed at the monitor as simple audible or visual alarms or in more sophisticated systems can be integrated into map or digital displays or pen traces.

SUMMARY

Battlefield surveillance is a vital part of the process of obtaining the combat intelligence required by a commander for the planning and conduct of operations; it is not an end in itself. An effective target acquisition capability is necessary if a commander is to make best use of the weapons available to him. The depth to which his surveillance and target acquisition system must be effective are dictated by the depths of his areas of interest and influence. He requires a 24 hour capability in all weathers which makes a use of several complementary techniques essential. The requirement is technically demanding and financially expensive.

SELF TEST QUESTIONS

QUESTION 1 What is meant by surveillance?

Answer ....

....

....

....

QUESTION 2 What is the difference between a commander's Area of Interest and his Area of Influence?

Answer ....

....

....

....

....

....

QUESTION 3 What is the maximum range over which an observer could reasonably be expected to observe in North West Europe?

Answer ....

QUESTION 4 What is the main disadvantage in using an active system?

Answer ....

....

QUESTION 5 How can surveillance ranges be increased beyond line of sight?

Answer ....

....

....

QUESTION 6 Is radar essential in a surveillance system?

Answer ....

....

QUESTION 7 Will one type of sensor fill the requirement for a surveillance system?

Answer ....

....

....

....

QUESTION 8 Why is contrast important in surveillance and target acquisition?

Answer ....

....

....

....

QUESTION 9 Most surveillance systems use electro magnetic waves. Name 2 other types of emission which may be used.

Answer ....

....

QUESTION 10 Why is target location more difficult in indirect fire systems?

Answer ....

....

....

ANSWERS ON PAGE 187

The fovea is a small depression on the retina of about 1.5 mm diameter and is the region on which light from a distant object on the optical axis is focused. It contains only cones and this gives it the ability to perceive fine detail and colour vision under daylight conditions though it is inoperative for night vision. As the angle from the fovea increases the concentration of cones rapidly decreases until at angles greater than 20 degrees from the fovea the only photo-receptors are rods and these are the main sensors for visual operation at night.

As can be seen from Fig. 2.1 the retina has a pronounced curvature. Light which enters the eye at a large angle of incidence to the optic axis is focused on a point on the retina almost on the hemispherical boundary and in a region which is totally dominated by rods. This peripheral vision is usually followed by a fixation reflex in which the eye moves to bring the image of the object onto the cone rich fovea for detailed viewing.

There are about 1×10^8 rods but only 6×10^6 cones and 1×10^6 nerve fibres. Hence some of the nerve fibres are connected to very large groups of rods which act to sum the effects of light over a comparatively large area of the retina. This integration of light over many rods improves the signal to noise ratio under low light conditions and, to some degree, compensates for the reduced ability to perceive detail. Only a small proportion of cones are connected by a single line to the brain, these being confined to the fovea region and explain the high degree of visual perception when this region of the retina is employed.

Photopic Vision

Daylight vision is also known as photopic vision and there is an associated photopic retina which is composed almost entirely from cones. These are activated by luminance or brightness levels above 3×10^{-2} candela m^{-2} and the eye is fully light adapted at one hundred times this threshold value. The upper limit of visual tolerance occurs for luminance levels of about 3×10^5 candela m^{-2}. The photopic retina has the capability of colour vision.

Scotopic Vision

Scotopic vision is associated with rod response and dominates at low light levels. The rods are actuated by luminance levels below 3×10^{-2} candela m^{-2} and the eye is fully dark adapted at a luminance of 3×10^{-5} candela m^{-2} to the lowest level of seeing which occurs at 3×10^{-7} candela m^{-2}. The time taken for full dark adaption depends upon the change in lighting conditions but is typically about 30 minutes from high level photopic vision to scotopic threshold. This is due to the time taken to build up a sufficient concentration of the chemical rhodopsin in the rods. The rhodopsin is quickly bleached by exposure to light leading to stimulation of the optical nerve. This stimulus depends only upon the quantity of light which is absorbed by the rhodopsin, not the type of light, and is the reason why there is no discrimination of wavelength in the scotopic retina, and why we do not perceive colours at night.

Dark Adaptation

The general features of dark adaptation are shown in Fig. 2.2 which plots the threshold of visual stimulation against the time from which the eye is exposed to a preadapting light.

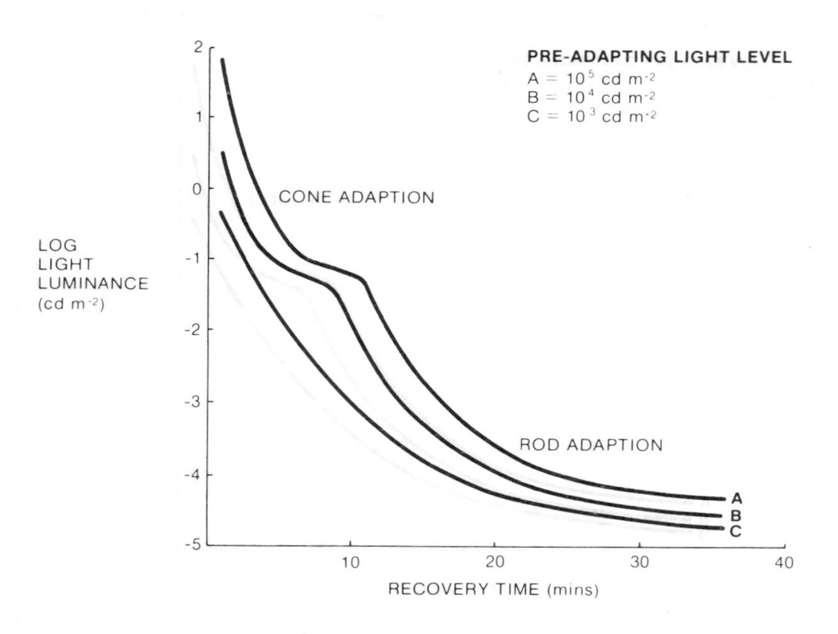

Fig. 2.2 Dark adaptation

From this figure it can be seen that although both rods and cones begin to adapt immediately, it is the latter which determine the vision threshold for short recovery times following exposure to a preadapting light source. However the cones quickly reach saturation threshold and it is the rods which continue to adapt for longer times. It can also be seen that as the level of preadapting illumination falls, the quicker is the process of dark adaptation, and the less evident is the two stage process until at the lowest levels recovery is completely controlled by rod adaptation.

The data in Fig. 2.2 actually refers to full preadaptation to white light and to a blue test threshold flash light of controllable intensity. Similar dark adaptation characteristics apply when both the preadapting light and the test light have different wavelengths. In general the dark adaptation process is quicker when the preadapting wavelength is longer and the test wavelength is shorter.

This suggests that to preserve dark adaptation under battlefield conditions red goggles should be worn. However the threshold luminance level for red (test) light is about 100 times greater than for blue light and one way of increasing the sensitivity of the dark adapted eye is to illuminate dimly lit areas with blue light. This has the further advantage of maintaining concealment because blue light occurs in natural phosphorescence, in woodland and sea environments in particular.

In the case of exposure to a pulse of light such as a gun flash the initial rate of recovery is expected to be rather quicker than for a steady light source of the same luminance but the later stages follow the same slower process.

The discussion above has been related to the condition where the object is merely either seen or not seen. If detail is to be perceived, the process of dark adaptation will take longer and the terminal threshold will be higher, to a degree depending upon the level of visual acuity to be achieved. For example the threshold for reaching the standard visual acuity of 1 minute of arc is 3 candela m^{-2} which is four orders of magnitude higher than for mere detection of a visual signal.

Rate of dark adaptation is also determined by the contrast in the scene as Fig. 2.3 illustrates.

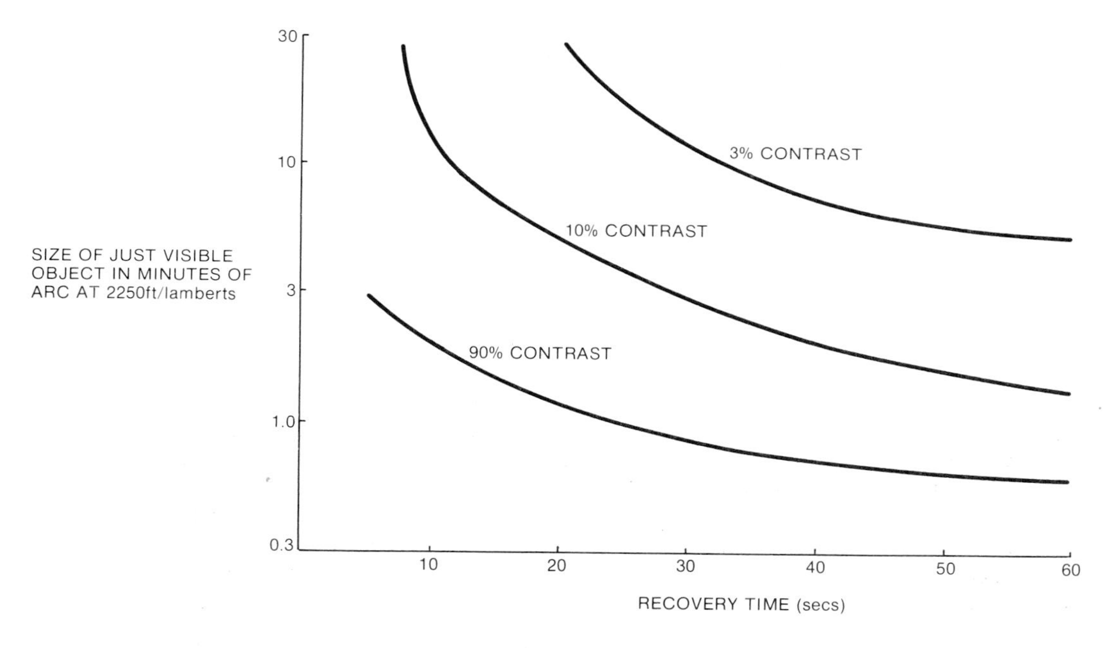

Fig. 2.3 Dark adaptation as a function of contrast ratio

Light Adaptation

The process of light adaptation is quicker than dark adaptation and sensitivity recovers completely in about 10 minutes. Glare reduces perception of detail because light scattering within the eye reduces the brightness contrast ratio. However the eye can detect almost at once an object only 1% as bright as the object which has just been looked at, as is commonly experienced by driving against headlights along country roads at night.

Mesopic Vision

Within the intermediate luminance region extending from about 10^{-3} candela m^{-2} to 1 candela m^{-2} vision is controlled by both rods and cones, within the middle region of the retina. This mesopic retina is responsible for twilight vision and in particular for the process called Purkinje phenomenum. The explanation is facilitated by Fig. 2.4 which shows the relative wavelength response of rods and cones.

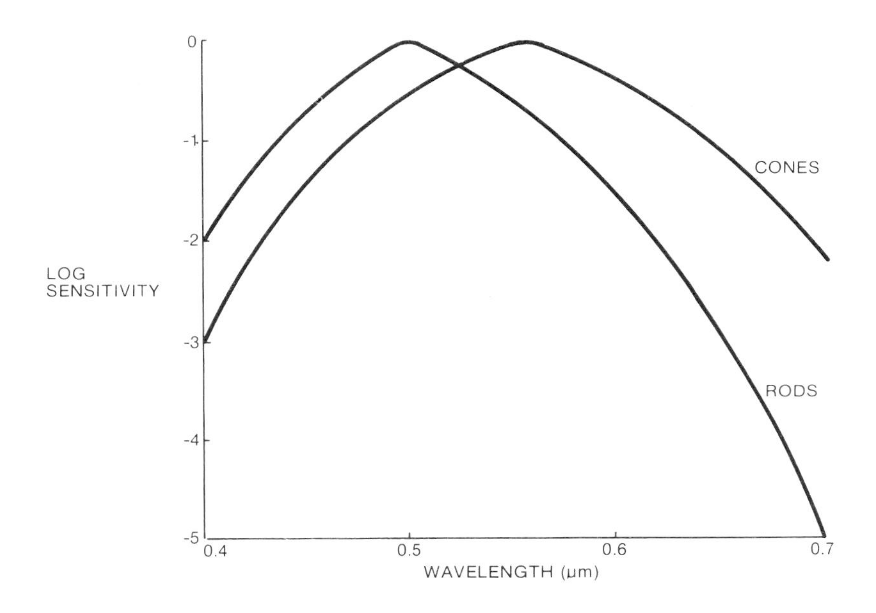

Fig. 2.4 Wavelength response of rods and cones

It can be seen that the maximum wavelength sensitivity shifts from 0.55 μm under daylight viewing to 0.5 μm for the dark adapted eye. Thus the brightness and hues of colours which are apparent in normal light conditions are different under twilight viewing and the dark adapted eye is much less sensitive to red light and more sensitive at the blue end compared with the light adapted eye. The change in the maximum sensitivity is known as the Purkinje shift and accounts for the apparent colour changes which occur in the twilight region. Thus as the light fades the relative luminances change and a red object which appears brighter in daylight than an adjoining blue object becomes darker as the level of illumination falls.

When objects lose their colours they become more difficult to recognise. The changes in brightness also makes shadows harder to dissociate from real objects and these apparent shadings make distances difficult to judge because three-dimensional vision concepts are disturbed.

Pupil Reflex

Figure 2.5 shows the changes which occur to the iris on the application and removal of a 320 candela m^{-2} luminance. The light adapted eye takes about half an hour for the pupil to dilate but only 10 seconds to contract again when exposed to a moderate light level.

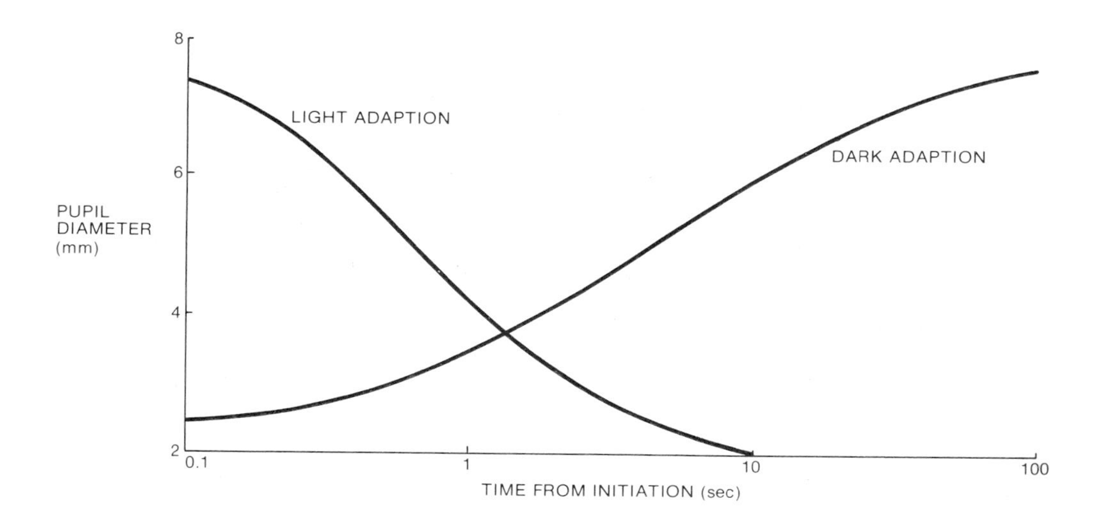

Fig. 2.5 Pupil reflex

Pupil reflex tends to optimise visual performance. At low light levels where photon noise and absolute signal level are the primary limits on perception, the pupil enlarges to increase the light gathering power. At high light levels, where the imaging performance of the eye is the controlling factor, the pupil contracts to give an optimum balance between diffraction effects and aberrations. It also reduces retinal illumination and consequently speeds up dark adaptation.

Visual Acuity

The visual acuity of the eye is a measure of its ability to perceive detail in a scene and is the opthalmic analogue of the term 'resolution' which is used to define the spatial threshold performance of an optical instrument. It is defined as the reciprocal of the minimum angle subtended at the eye by the resolved object detail. Standard test objects may be a single line or a set of black and white bars of decreasing separation, but other shapes and letters are commonly employed in clinical tests. The minimum angle by which the object can just be resolved, that is the limiting line width in a single bar or two adjacent lines in a bar pattern, is called the visual angle. The visual angle is usually measured in minutes of arc, thus the visual acuity has units of reciprocal minutes, or cycles per minute when referred to the just-resolved line frequency of the bar pattern.

Visual acuity varies with the luminance or brightness level in the scene as Fig. 2.6 shows.

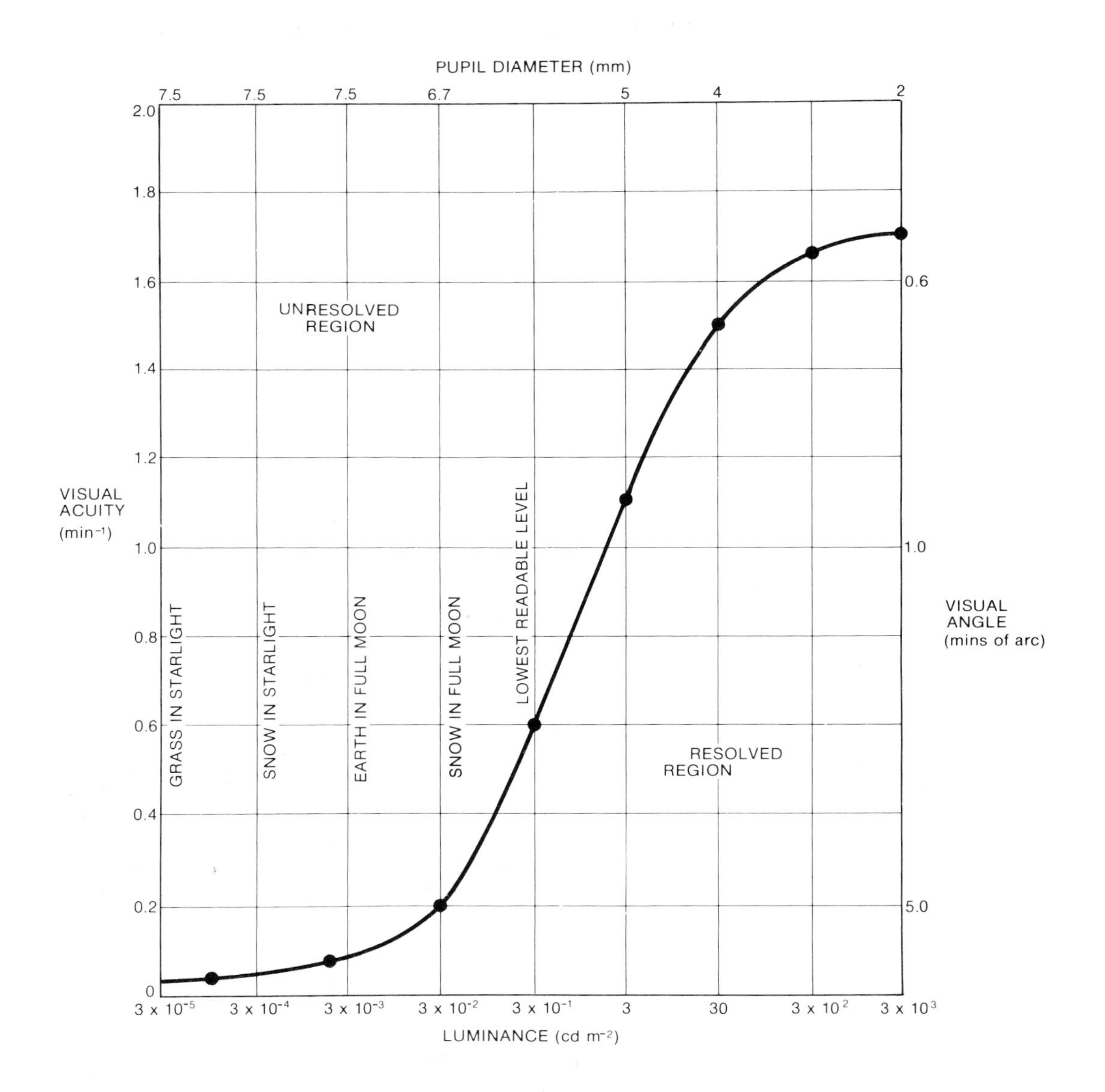

Fig. 2.6 Variation of acuity of the eye and pupil diameter

From Fig. 2.6 it can be seen that the visual acuity increases with the luminance level from about 0.02 min^{-1} at the threshold of vision to a maximum resolution of 1.7 min^{-1} at a luminance of 3×10^3 candela m^{-2}. The associated visual angles are 50 minutes and 0.6 minutes corresponding to a minimum perceived object size, at a range of 1 km, of 15 m and 0.15 m respectively. As long as some light is seen some form of vision remains and even on moonless nights it is usually possible to distinguish sky from ground although the visual acuity is very low. The visual acuity of the normal eye in clinical tests is taken to be 1 min^{-1} at a specified light level and contrast.

Visual acuity varies with the contrast in the object field as Fig. 2.7 illustrates.

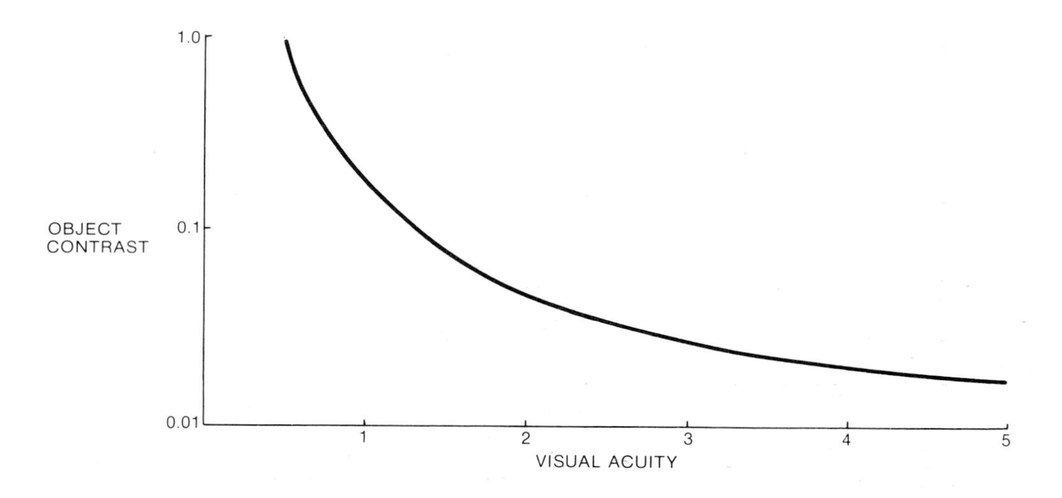

Fig. 2.7 Variation of visual acuity with object contrast

A further important factor is the diffraction effects which occur because of the finite size of the eye pupil, and these dominate for bright conditions when the pupil is smallest. As with any other optical detector the graininess or photo-receptor density on the retina will control the maximum resolution. If the separation between a pair of images is to be perceived a receptor must be present in the intervening space. Since the diffraction effects tend to diffuse the edges of the image the separation may not be clear cut resulting in a "grey" zone between the two images which the retina may or may not resolve as a separation.

The Rayleigh criteria for defining resolution is discussed in the optical instruments section but also applies to the eye.

For the dark adapted eye the pupil is fully expanded and defocusing will occur from spherical and chromatic aberrations leading to a reduction in visual acuity. Overall the optimum visual acuity occurs at a pupil diameter of about 2.5 mm.

Visual acuity varies over the retina and for daylight conditions is greatest in the central region particularly the fovea where the cone photo-receptors are closely packed. At 40 degrees to either side the visual acuity falls to only 5% of the foveal value, thus small details in the peripheral field of view have to be about 20 times larger in order to see them with the same clarity.

At low illumination the iris opens wider and the rods dominate. The eye becomes colour blind and the fovea becomes the blind spot since the cones lack the sensitivity to respond.

Vernier Acuity

Vernier acuity is the ability of the eye to align two objects such as two lines or cross-wires. The eye is very capable in this operation giving repeatable performance to better than 5 seconds of arc, thus its vernier acuity is an order of magnitude better than its visual acuity. The vernier acuity is best when setting one line between two, closely followed by setting a line on cross-wires or aligning two butting lines, but less effective at superimposing two lines. These properties are taken into account in designing coincidence range finders and sighting marks.

The narrowest black line on a bright background that the eye can detect is about 1 second of arc, and this is borne out by the ease with which a whip antenna can be seen at considerable distances.

The eye can detect motion to the accuracy of about 10 seconds of arc. The slowest motion detectable is about 1 minute of arc per second and the fastest about 200 degrees per second.

Light Contrast

The eye is a poor photometer in assessing the absolute level of brightness but is an excellent comparator of brightness or colour tones. Figure 2. 8 indicates the minimum brightness difference that the eye can detect at a function of the luminance of the test area.

Stereoscopic Vision and Illusions

Stereoscopic or binocular vision is a manifestation of the interaction between the two eyes and the brain. It is a developed perception arising from ciliary muscle adjustment and convergence, together with acquired environmental perspective. There is no neurophysiological explanation of stereoscopic vision and the attainment of single vision may be due to physiological fusion, in the cerebal cortex, of image pairs arising from corresponding regions of the two retina. Both left and right sides of each retina are connected to each other via optic tracts which remain separate until they reach the cortex where three-dimensional fusion occurs. This process is a prerequisite for good binocular vision and is the basis for the design of modern prismatic binoculars.

In monocular vision the estimation of distance and relief of objects is very imperfect. With binocular vision however convergence should make depth estimation unambiguous but illusions do occur as is evident by well known geometrical examples of depth inversion, eg Neckers cube.

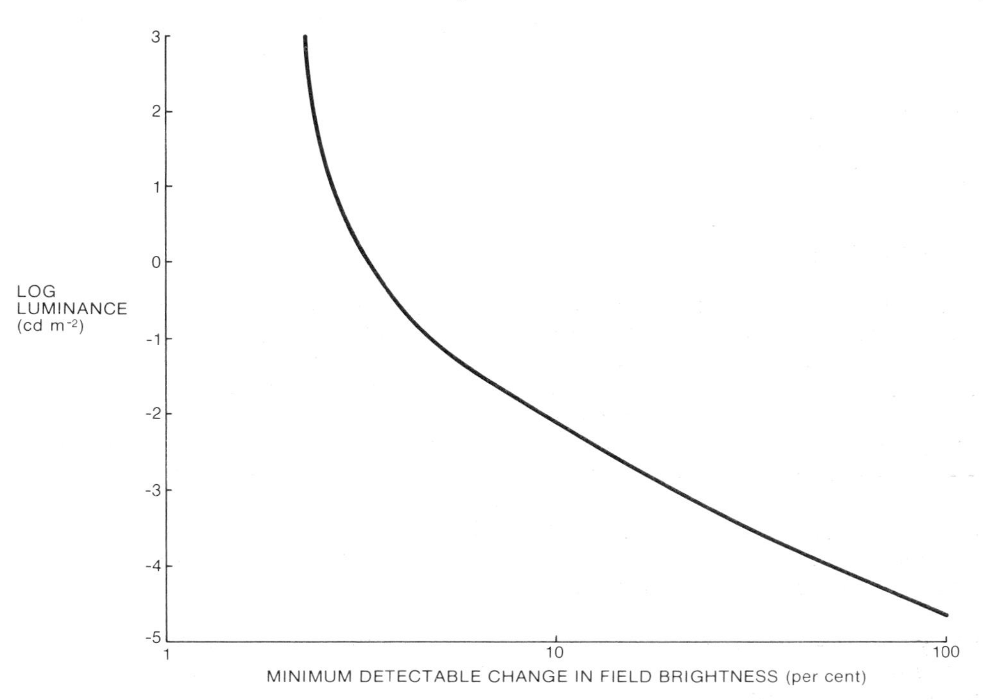

Fig. 2.8 Light contrast capability of the eye

Other examples have clear military significance. A single fixed point of light against a dark background appears to move randomly: moving the eyes and avoiding steady fixation, or using blinking lights as in the case of an aircraft, helps to destroy this effect which is called Autokinetic Illusion.

Oculogyral illusion is caused by rotary acceleration. In this case a spot of light against a dark background will seem to move in the direction of rotation whilst acceleration is taking place but the movement will cease when constant velocity is attained. Upon deceleration the spot of light will appear to move in the opposite direction.

Oculogravic illusion is caused by linear accelerations and, like oculogyral illusion, is caused by stimulation of the vestibular sense organs in the auditory apparatus, which are responsible for posture and control. The effect can cause illusions in reading the illuminated dials in an aircraft and, if the pilot accelerates on overshoot on trying to land, can give him the impression that the nose of the aircraft is rising which is confirmed, falsely, by the artificial horizon when in fact the aircraft is still in descent.

Further visual illusory phenomena occur for a flashing light. There is a critical frequency at which the sensation of flicker ceases and the light appears to be continuous. This is called the critical fusion frequency and varies with the light intensity and object size. Usually this occurs between 20 and 50 per second and is the basis of display systems like television. Flicker evokes responses in the cortex and can cause anxiety or giddiness, producing a form of epilepsy in susceptible people.

When two or more stationary lights which are close together are emitted with a delay between them they can be perceived as a moving light, the principle of which is used in some displays.

Visual Search Time

The time to search a stationary image display depends upon the fixation time of the eye, which is approximately 0.3 sec, and the angular size of the display. The circular field of clear vision is usually taken as 10 degrees so that the search time is computed by determining the number of non-overlapping 10 degree fixations required to cover the field. Thus coverage of a 16 degree x 12 degree scene would require a search time of about one sec. Longer times may be required in conditions of low contrast.

In searching a moving display the eye has to jump about in order to see all detail adequately in the foveal field, but unwanted objects often mask the wanted ones. The eyes have to move at exactly the same rate as the display otherwise the image of an object moves over the retina, blurring the image and making fine details harder to see with increasing difficulties as the display moves faster. Although observers with good peripheral acuity do better at slow display speeds it is not of much advantage at high speeds where good central acuity is best.

Such effects are important in television displays and a compromise has to be reached between including all relevant information and the rate of scanning the face of the tube. Ideally a system should only give information when needed.

OPTICAL INSTRUMENT DESIGN

Introduction

The eye has a remarkable dynamic response range of about 10^{12}:1 which is far greater than any other single light sensitive device. Nevertheless the ability to see detail in a scene becomes worse as the luminance decreases until only large objects can be discerned, but not clearly recognised. Because sensitivity for the dark adapted eye is lower, peripheral vision, using rods, is more effective than foveal vision, using cones, under low light conditions and this has an important bearing on visual perception in night surveillance operations.

The purpose of an optical instrument is to improve the performance of the eye by making sighting more precise and by improving the detail perceived in the observed scene especially under low light conditions. Some magnification of the object makes targets more easily identifiable but field of view is usually reduced in proportion, aiming errors are increased and, if the magnification exceeds the diffraction limit, no further enhancement of detail is possible. In fact an optical sight of unity power has advantages over the unaided eye in that it superimposes a sighting mark exactly on the image plane and thus eliminates the necessity of trying to aim a foresight on a distant target through a backsight aperture.

In any optical instrument design the magnification has to be considered with all other interdependent factors which often lead to a compromise. In general relatively lower magnifications and larger fields of view are used for surveillance instruments such as binoculars whilst relatively higher magnifications and smaller fields of view are necessary for target acquisition.

Optimum Magnification Requirements

The diameter of the objective lens in an optical system can be made much larger than the pupil of the eye and this results in greater light gathering power and improved resolution. In the latter respect the limiting angle of resolution α is given by the Rayleigh criterion

$$\alpha = 1.22 \frac{\lambda}{D}$$

where α is the minimum angle subtended at the instrument by two distant point objects to discern them as separate and equates to the visual angle for the special case of the eye; D is the diameter of the objective lens and λ is the wavelength of the light. Typically for $D = 50$ mm and $\lambda = 0.55\,\mu$m,

$$\alpha = 1.36\ 10^{-2}\ \text{mrad}$$

$$= 5 \times 10^{-2}\ \text{min}$$

This is approximately one twentieth of the visual angle of the eye under daylight illumination and to make full use of this better resolution the optimum magnification of the system should be 20. Less magnification will result in loss of resolution whilst greater magnification will only magnify the coarseness and further reduce the brightness of the image.

It would be impossible to match the eye resolution for low light conditions because this would require enormous apertures and magnifications and in practice military optical instruments never operate at their diffraction limit at night-time.

The effect on visual acuity of a (7 x 50) night binocular is shown in Fig. 2.9 overleaf.

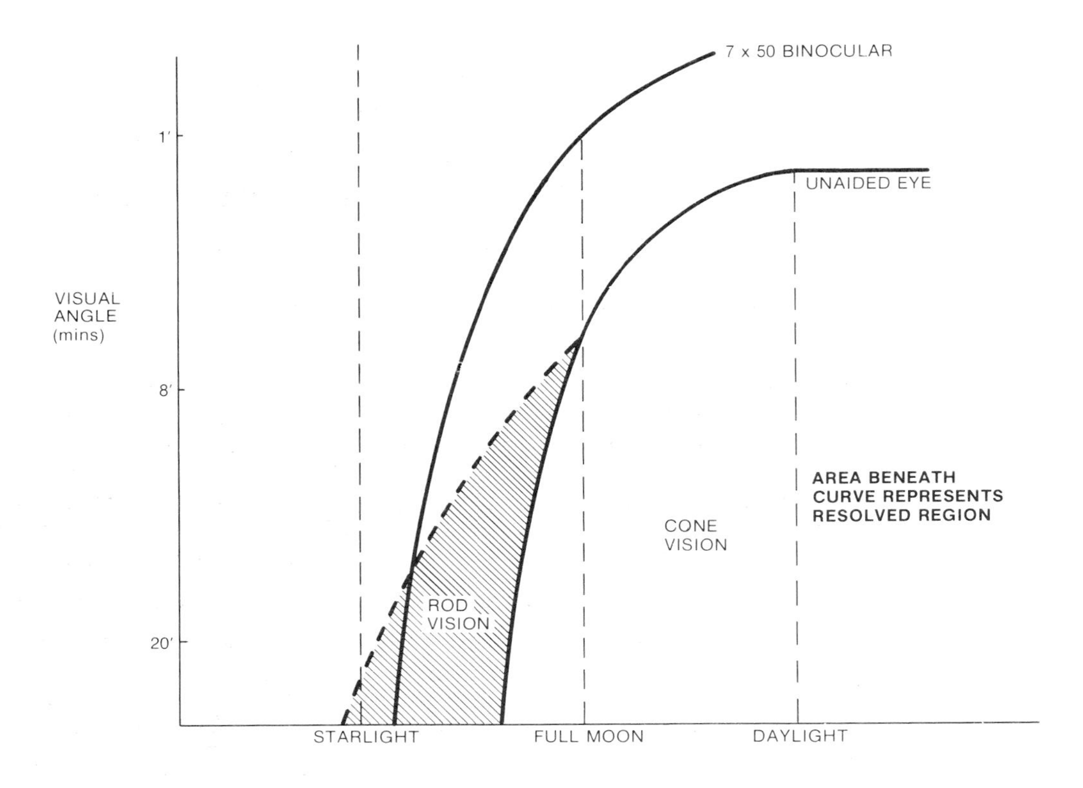

Fig. 2.9 Improvement of eye performance at low light levels

The diameter of the objective is much larger than the pupil of the dark adapted eye and therefore collects more light from the scene. These produce a magnified image on the retina which in effect shifts the visual acuity curve towards smaller values of visual angle α by a factor nearly equal to the magnification. A firmly mounted (7 x 50) binocular increases the effective visual acuity of the eye by a factor of approximately 7 and improves the light gathering power by a factor 50 to maintain the same image brightness. This instrument extends the range of military surveillance down to light levels of approximately one quarter of moonlight and gives a dusk to dawn operational role. Hand tremor reduces the detail seen in the image to that which could be perceived by a firmly mounted x4 binocular. In practice hand held binoculars with magnifications exceeding x10 become very difficult to use efficiently.

As we have seen the visibility of a target is also influenced by field luminance, by contrast between target and background, and by the complexity of the scene. Elongated objects such as a gun barrel or a whip antenna seen against the sky or a well lit background may be resolved even when the thickness of the elongated target subtends a much smaller angle than that determined by the diffraction limit. However an object in a complex field of view may have to subtend an angle of an order of magnitude greater than the diffraction limiting angular resolution before it can be resolved as such.

Optical Depth of Field

Diffraction effects in optical systems have another consequence on their performance, namely the optical depth of field.

A point source is imaged as a diffraction disc of radius r where

$$r = 1.22 \frac{\lambda f}{D}$$

where f is the focal length of the objective lens of diameter D and λ is the wavelength of the light.

The overlapping of diffraction discs on the images from different objects gives rise to an image focal depth which allows objects lying at ranges either side of the optimum focus range to be regarded as being in reasonable focus in the image plane. The limits in range over which the objects are acceptably in focus is known as the depth of field. Since the effect depends upon diffraction it follows that the depth of focus will be longer for small values of D when diffraction effects are most pronounced. More specifically since it depends upon the value of r, which can be written as $r = \text{constant} \times \frac{f}{D}$ then a greater depth of field is achieved for greater values of $\frac{f}{D}$, or f-number, stop-number, or relative aperture of the system to name three alternative forms for this ratio. In practice a lens is usually specified by its f-number F4, F8, etc corresponding to $\frac{f}{D} = 4$, and $\frac{f}{D} = 8$ respectively. The depth of focus is greater at F8 than at F4 but only at the expense of the field of view. Hence a compromise is necessary to balance the conflicting requirements of surveillance and target recognition. Since the latter requires fine detail and therefore high resolution a large aperture lens is essential. This limits the depth of focus and increases the closest in-focus range. Surveillance on the other hand requires a large field of view and a relatively low magnification.

In night vision systems employing a fixed focus, long range focusing is undesirable because of the limitation of photon noise. Focus adjustment is frequently provided to deal with shorter ranges but this increases the complexity and expense of the equipment.

Parameters of an Optical System

The main features of an optical system are set out in Fig. 2.10. The aperture stop is the limiting factor which determines the amount of light from the object transmitted through the system. The entrance and exit pupils of the system are the images of the aperture stop formed by all elements in the object and image space respectively. Thus the entrance pupil lies within the objective lens, which controls both the amount of light gathered and the resolution, and the exit pupil lies in the plane at which the principal ray crosses the axis to the right of the eye lens. The distance from the eyepiece lens surface nearest the eye to the exit pupil is called the eye relief. For maximum sensitivity the eye is placed at the exit pupil which should be the same size as the pupil of the eye.

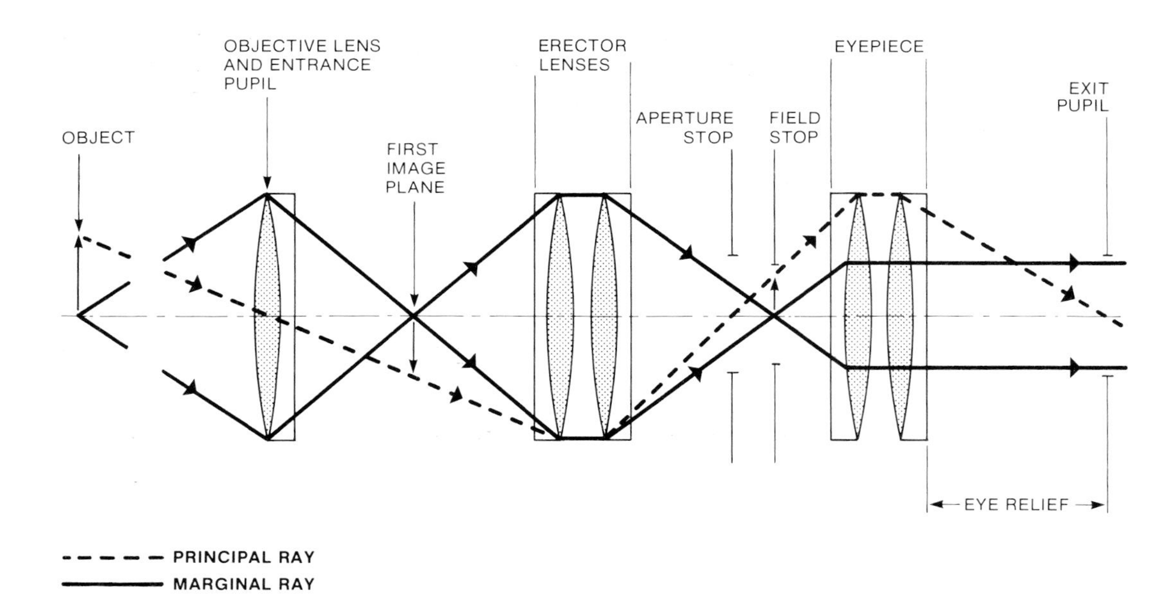

Fig. 2.10 Main parameters of an optical system

The field stop limits the field of view since this diaphragm would prevent a principal ray starting from a point on the object that is further from the axis from passing through the system. Other types of stop are used, for instance a glare stop to cut out stray light and baffles to remove multiple scattering within the system.

The importance of these optical parameters will become manifest in the section which deals with military surveillance instruments.

SYSTEM EVALUATION

Bar Pattern Analysis

As mentioned in the discussion on the visual acuity of the eye, the determination of limiting resolution is usually achieved by presenting the observer with a bar chart consisting of a series of alternating black and white lines in which the width of the black lines equals that of the white lines, progressively making the separation and line width smaller until a limit is reached at which the observer just cannot resolve a pair of adjacent lines as separate. This line limit expressed as line pairs per mm or cycles per mm represents the spatial frequency at which the observer's eye can no longer discriminate differences in the light and dark transitions in the bar image. Exactly the same considerations apply when an optical instrument is used.

The capability to perceive single military targets can be expressed as a function of the limiting resolution per target minimum dimension. Tests on the recognition of a main battle tank of minimum dimension 2.3 m show that for 50% probability of being correct:

1. To detect a target requires 1.0 ± 0.25 line-pairs per target minimum dimension, or a mean of 2 lines.

2. To orientate the target requires 1.4 ± 0.35 line-pairs per target minimum dimension, or a mean of 2.8 lines.

3. To recognise the type of target requires 4.0 ± 0.8 line-pairs per target minimum dimension, or a mean of 8 lines.

4. To identify the target requires 6.4 ± 1.5 line-pairs per target minimum dimension, or a mean of 12.8 lines.

The concepts are illustrated in Fig. 2.11.

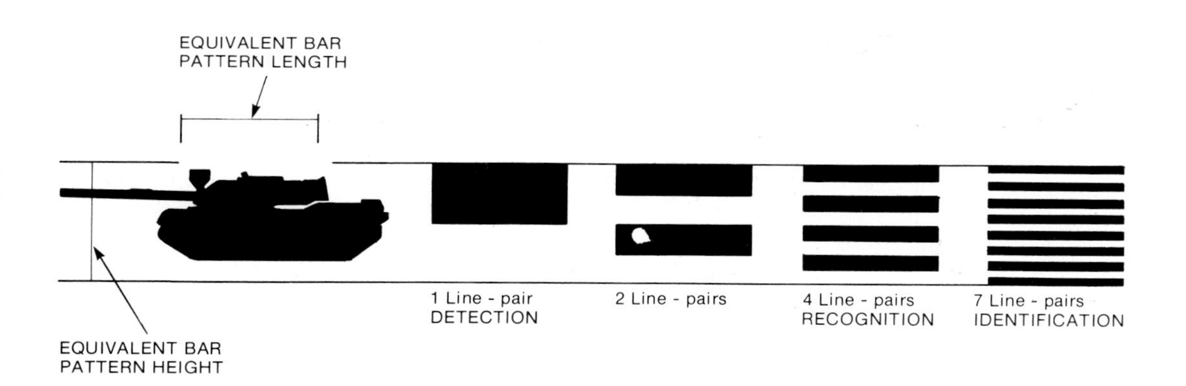

Fig. 2.11 Bar pattern analysis criteria

There is a further criterion specifying detection and recognition in 2 dimensions. Scenes can be synthesised in terms of black and white dots, the number of dots defining the quality of the picture. For example, a picture of a Land Rover in an image display having a 4 x 3 aspect ratio required 500 picture dots for detection, 1300 picture dots for shape and orientation recognition, and 5000 picture dots for recognition of the target. Detail in excess of 20,000 dots is rarely required, at least in battlefield conditions.

Optical Transfer Function (OTF)

These simple concepts of pattern identification and recognition are extended to the quantitative evaluation of optical and electro-optical system performance in a manner analogous to the evaluation of the performance of an electronic amplifier. The input to the optical equipment is a bar chart of varying contrast and spatial frequency, and the output is also assessed in terms of contrast and spatial frequency. The ratio of the output to the input is known as the Optical Transfer Function (OTF) and it serves to indicate the performance of the system as a function of frequency. In the case of the OTF for the eye the contrast sensitivity maximises at 2 cycles per degree falling off at lower frequencies because of ocular tremor and at high frequencies due to aberrations, diffraction and finite receptor size, see Fig. 2.12.

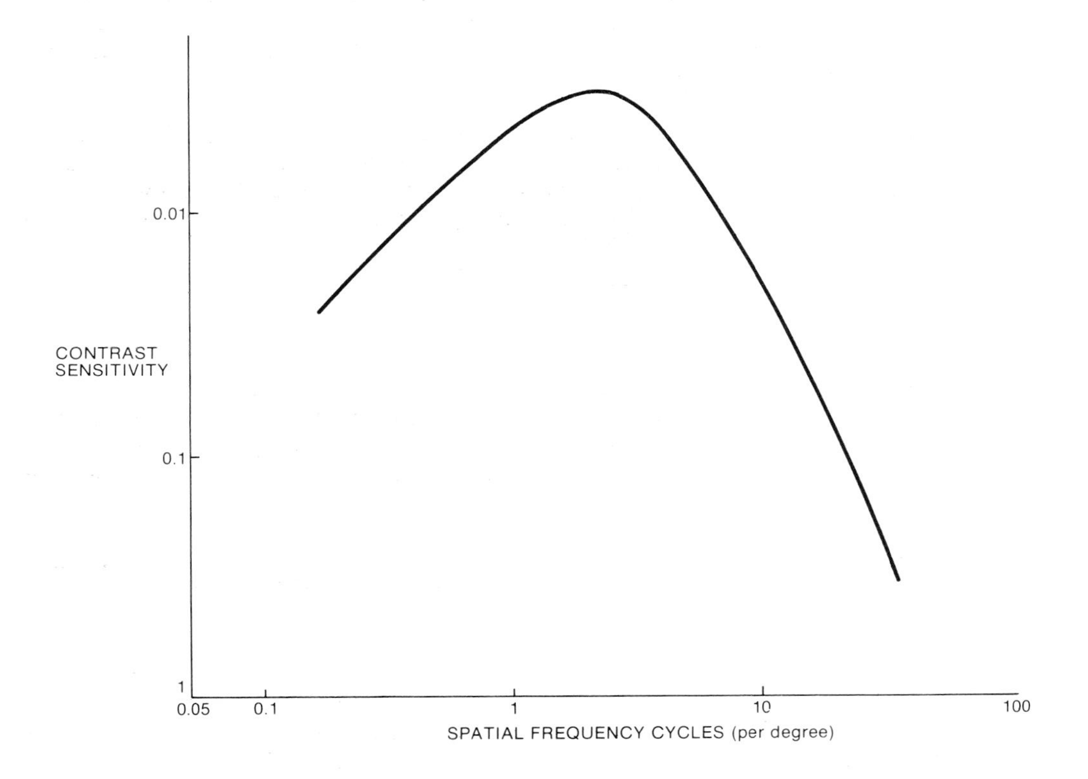

Fig. 2.12 Optical Transfer Function for human eye

MILITARY SIGHTING AND SURVEILLANCE INSTRUMENTS

Open Sight

The oldest form of aiming is the open sight, where a foresight is situated at the muzzle and the backsight is close to the eye. The backsight is shaped in the form of a U, V, or ring and the foresight is usually wedgeshaped but can be circular. The arrangements make use of the eye's vernier acuity.

Though the foresight is often illuminated by a beta light the open sight is difficult to use in low light levels and is relatively slow because there is little eye relief and it is not easy to follow a target without losing aiming accuracy.

Historically guns have been designed to keep the minimum of the firer's head exposed to counterfire and also to align the sight close to the bullet trajectory. The resulting arrangement produces an upward recoil on firing and large lateral accelerations are set up in the weapon which can cause significant aiming errors.

Recent straight-through guns eg EM2 and M16 have been developed in which the barrel is in line with the shoulder and the jump of the weapon is much reduced. The high sight line produces a rather complicated foresight. Thus the need for greater accuracy and protection results in a rather untidy arrangement and considerable training is required to use it properly.

Open night sights are in service with the British Army and NATO. These are designed to enable accurate aiming at ill-defined targets in low light levels and poor visibility. There are several adaptations but all work on the principle of illuminating the iron foresight. The tritium light source requires no power supply and has an operational life of at least 10 years.

Telescopic Sight

Improved sights are an adaptation of a conventional optical telescope fitted with a graticule, sometimes called the reticle, that can be illuminated. Some magnification is provided to make optimum use of the eye. The exit pupil is designed to match the dark adapted eye and so improve low light performance. A suitable eye relief is chosen to protect the firer's eye from recoil injury.

The principal components of the small arms sight are shown in Fig. 2.13. This sight has a magnification of about 3x and is an adaptation of the basic terrestial telescope design.

The aperture stop in this and most other similar telescopes is the object glass, although it can be a diaphragm or image of a lens or diaphragm within the telescope. The exit pupil is the image of an erector lens or stop having the smallest angular diameter seen from the eyepiece. When the eye is placed at the exit pupil the full field of view is seen. A field stop forms a sharp edge to the field of view and could be on the 1st or 2nd image plane as is the sighting mark. This is the graticule or reticle, and is normally etched onto a piece of glass, but it can be cross-wires within a metal diaphragm.

Small arms sights of prismatic construction similar to half a binocular are now made in quantity for the British Army: they have remarkedly good light transmission and high optical quality.

Fig. 2.13 Small arms sight

An example of an infantry sight now in service for the British Army is the TRILUX SUIT L2A1 which was developed by RARDE. It is composed of an optical telescope with prismatic lateral shift for user comfort. It has a magnification of 4x, a field of view of 8 degrees, an eye relief of 35 mm with a rubber eyeguard to reduce external light. The exit pupil of 6 mm matches the dark adapted eye. Within the field of view the firer sees an aiming mark instead of a conventional cross-wire graticule; this pointer is illuminated with red light from a tritium source. The intensity of illumination can be varied to suit the ambient conditions by an external control. A two-position liner gives a choice of range adjustment between 0-400 m and 400-600 m. The compact system is designed to improve infantry weapon effectiveness in night fighting and increases the range of operation beyond conventional iron sights.

Larger sighting telescopes for the bigger guns evolved in the early part of the century and were in fact massively made terrestial telescopes often of rather low (3x) magnification with apertures up to 50 mm in diameter. The Army still uses sighting telescopes with low magnifications, but now they are smaller and iighter.

Long range artillery required greater magnification which could also be varied to suit the prevailing atmospheric conditions. This gave rise to the development of telescopes with variable power, and examples are to be found which have been used throughout the two world wars. Typical magnification ranges are 7x to 21x and these were forerunners of the modern zoom optical system.

Smaller periscopic 'Dial Sights' are used for indirect laying of artillery. These operate at a magnification of 4x and are of a rather complicated optical design. Due to the modern lens coatings later dial sights have much clearer and brighter images than the earlier ones.

Collimator Sight

The collimator sight provides rapid aiming and is fully developed in the single-point gun sight. The principle is that the graticule image is placed at infinity and thus it acts as a foresight at the target. This means that the firer has only to place this on the target and pull the trigger. In the single-point sight the graticule is viewed by the right eye whilst the target is viewed by the left eye. It is important for both eyes to focus naturally on the target but a uniocular version has been constructed whereby a beam splitter superimposes the graticule on the target view. Eye relief is not a problem and eye position does not materially affect the aiming accuracy thus the collimator sight has some advantages which make it simple to use.

Prismatic Binocular

The layout of a typical prismatic binocular is shown in Fig. 2.14.

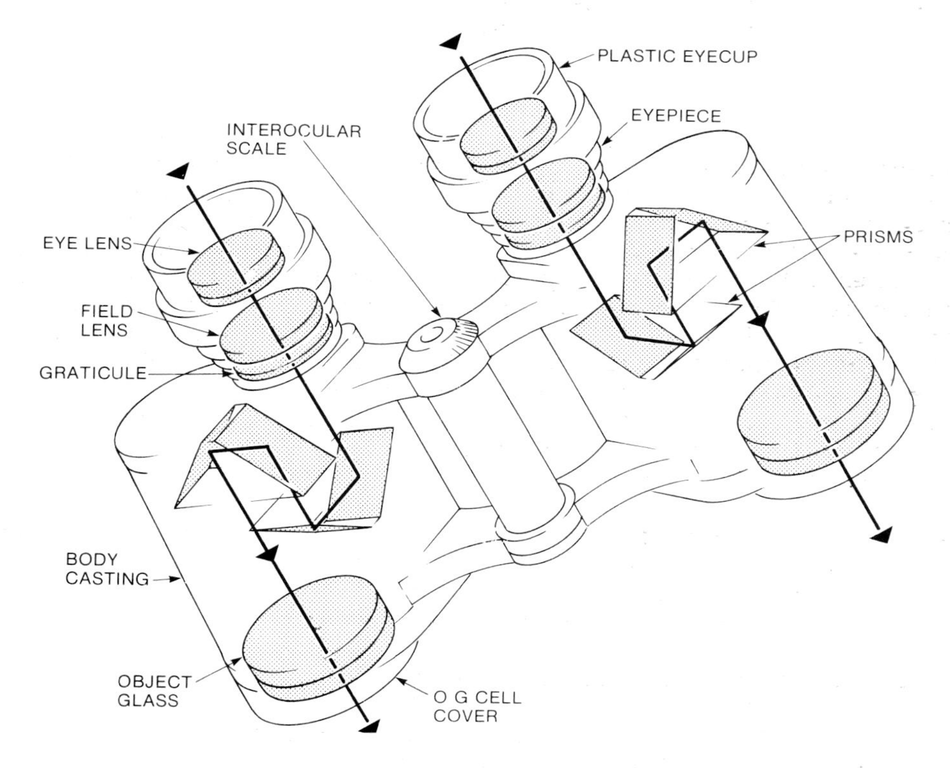

Fig. 2.14 Prismatic binocular

The object glass is the aperture stop and entrance pupil. Two prisms in each optical path erect the image and give the characteristic shape to the binocular body. In the case of the 6x binocular shown the field of view is about $8\frac{1}{2}$ degrees. Each eye perceives a slightly different view and this produces the stereoscopic effect which is enhanced by higher magnifications and wider objective glass separation. The two optical members are mounted on a central hinge which can be adjusted to suit the eye separation.

Binoculars are usually described by two numbers: the first gives the magnification and the second the objective glass diameter. A 7 x 50 binocular has a magnification of 7x and a 50 mm diameter objective glass. Magnifications from 3x to about 10x are practical for hand held devices, higher magnifications emphasise hand tremor to an unacceptable degree. For specialist observation purposes large tripod mounted binoculars of 15 x 80 and 25 x 105 etc have been made and are still extensively used.

For a particular system the field of view is inversely proportional to the magnification although more complex eyepieces with aspherical surfaces can be designed to maintain the field of view at a higher magnification. In fact these complicated eyepieces can be designed for a 10x magnification which retains the $8\frac{1}{2}$ degree field of view of a 6x instrument.

There are two descriptions of field of view. The real field of view is the angle between the extremities of the field seen through the binoculars. This angle is multiplied by the magnification to give the apparent field of view which is an important concept in eyepiece design. Taking some typical cases a simple 3-lens eyepiece has an apparent field of view of 50 degrees but a multi-element extra wide field eyepiece could increase this to nearly 90 degrees. It is possible to make the apparent field of view too large for the eye to scan comfortably and in practice it is found that the optimum value is about 65 degrees.

The exit pupil diameter is equal to the objective glass diameter divided by the magnification. The size of the exit pupil governs the amount of light entering the eye and is a most important design criterion as it governs the ambient light condition in which the instrument can best work. Ideally the exit pupil should be the same diameter as the eye pupil because under this condition the view seen through the binocular would have the same apparent brightness as seen by the naked eye. If the eye pupil is smaller than the exit pupil the naked eye brightness will be maintained but if the converse is true the image brightness will be reduced and this is undesirable in an instrument which is designed to make optimum use of poor light conditions.

From the information above it is possible to design suitable binoculars for particular uses as the table below illustrates.

Exit Pupil Diameters and Light Conditions

Up to 3 mm	Suitable only for bright conditions (6 x 18; 8 x 24; 10 x 30; etc).
3 to 5 mm	General purpose use in daylight (and twilight if nearer 5 mm).
5 to 8 mm	General purpose in daylight, twilight, at night and from moving vehicles (5 x 40; 7 x 50; 8 x 60).

The desirability of a large exit pupil when the binocular is being used in say a helicopter can be imagined, where it is difficult to hold the instrument exactly to the eye due to vibration. In these circumstances a daylight eye pupil of 3 mm diameter will remain within the 8 mm exit pupil of a 5 x 40 binocular even under the most difficult condition.

The position of the exit pupil is also an important factor. A long eye relief can limit the apparent field of view. In the case of a service type binocular eg 7 x 50 the eye relief is about 20 mm. The eye pupil should coincide with the exit pupil to make full use of the field of view and use the binocular to greatest effect.

An eye relief of 20 mm allows for rubber eye-cups that can be folded down to accommodate spectacle wearers. These eye-cups are designed to be in their normal position for use without spectacles so that the eye is held just on the exit pupil. When folded down they allow the spectacle wearer to maintain his eye pupil at the exit pupil and avoids the loss in field of view that would otherwise result. A long eye relief also allows eye protection to be worn.

A relatively recent development in optical instrument technology has been the advent of anti-reflection coatings to lens and prism surfaces. Each untreated air to glass surface reflects about 5% of the incident light but this can be reduced to about 0.15% with the latest multi-layer anti-reflection coating. The process is called blooming and such coatings are easily recognised by their reflected colours which are usually purple or green. In a normal 5-optical element telescope the overall light transmission of the instrument would be about 60%, the remainder being scattered to the detriment of image contrast. Blooming increases light transmission to 80% or more and makes identification and recognition of objects easier.

Military binoculars are designed to withstand rough treatment and are sealed to withstand ingress of damp. Central focusing adjustments are adequate for civilian use but are seldom robust enough for military environments. Military binoculars carry no central focusing adjustment, which is difficult to seal, but instead have independent eyepiece focus. Fixed focus has been in use for many years; when it is used, the instrument is set to the sight of an average user. This does away with the need to move focusing adjusters and makes maintenance and sealing easier. With sufficient eye relief normal spectacles can be worn to correct sight defects.

Even though the emphasis on military binoculars is towards robustness weight penalties are important and a well designed instrument should be lightweight, well balanced and easy to use. To this end 7 x 42 is now adopted as the ideal military binocular specification replacing the original 6 x 30s and 7 x 50s. So that the bulk can be reduced, modern roof prisms are used doing away with the traditional body shape. Long eye relief eyepieces are used for convenience and modern lens coatings and improved optical design gives crisp bright images.

Optical Range Finders

The optical range finder has largely been superseded by the laser range finder which is described in Chapter 5, but there are applications in which the passive technique is still deployed.

A coincidence range finder relies on the vernier acuity of the eye. It consists of two telescopes with reflectors in front of each objective lens and a common central eyepiece. The separate images are displayed above and below a central line. A correctly adjusted instrument would show a distant object as a continuous image equidistant about the horizontal centre line. If a nearer object is then viewed the two images will be laterally displaced. The images are brought into coincidence again by the lateral movement of a prism or reflector and the amount of correction required is calibrated as a measure of range.

The stereoscopic range finder exploits the stereoscopic acuity of the eyes and can be likened to a binocular with very widely spaced object glasses. This exaggerates the stereoscopic effect giving a positive three-dimensional aspect to all objects in the field of view. Ranging is done by virtue of a dot or a graticule in each eyepiece. If a dot is used it is fused stereoscopically and the rotation of a knob joined to a range scale causes an apparent movement of the dot until it is seen to be on the same plane as the target at which point the range is read off from the scale. If graticules are used the eye fuses them as a series of dots each marked with a number. The range is obtained by noting which dot is at a similar distance as the target and the associated number is converted to a direct reading of range.

A more unusual range finder works on the unequal magnification in the two optical members of the system. This is similar to the coincidence range finder in that two images are seen in a single eyepiece, one above and one below a central line. There is a range scale engraved on the central line and when the range finder is rotated slightly in azimuth the two images traverse the range scale. The speed of traverse depends upon the difference in magnification, one leg being typically 8x and the other leg 7. 5x. At some point along the centre line the two images coincide and the range is then read directly from the scale.

Stadiometric range finders are also used, these give an estimation of range by adjusting the diameter of a variable aperture until it just fits over a target of known size. The graticule in a binocular works in just the same way, only the subtence of the target in mils is estimated and range obtained from:-

$$\text{Range} = \frac{\text{Dimension of target (m) x 1000}}{\text{No of mils}} \quad \text{metres}$$

AFV Optics

AFV optics have evolved over the years from simple telescopes and periscopes to sophisticated and expensive multi-element systems. The use of efficient modern lens coatings technique permit all current military optical instrumentation to use more optical elements than would have been possible in the early years, as can be seen in Fig. 2.15. Prisms are usually preferred to mirrors for stability and ease of mounting, except when movement is required and a mirror is lighter to drive.

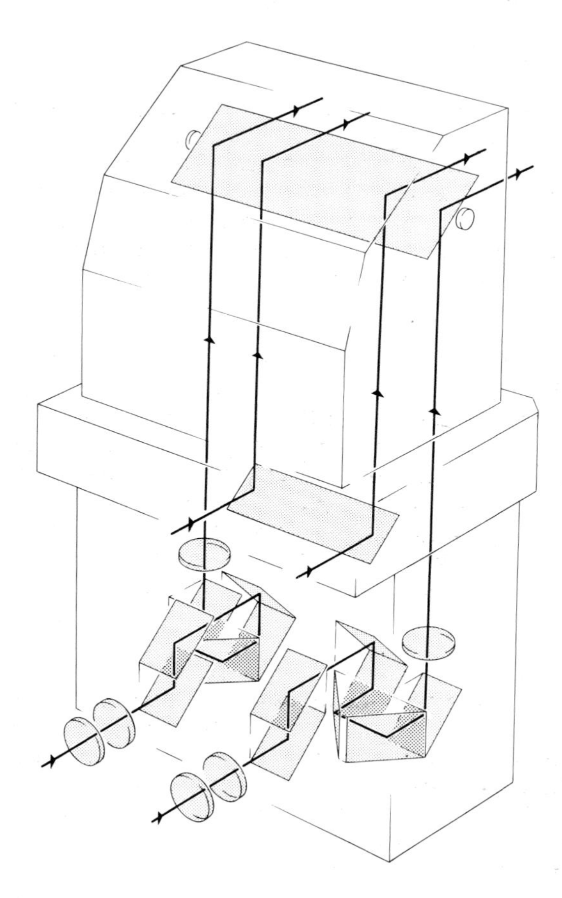

Fig. 2.15 AFV optics

A tank commander's periscope is usually provided with a low magnification (1x) channel for selecting a target and a high magnification (10x) for identification. An exit pupil of 5 mm at the high magnification is desirable and fields of view of 6° at 10x and 60° at 1x are a common choice.

FUTURE DEVELOPMENTS

Though there has been a major development in electro-optical sights over the past 15 years there will still be a need for a simple, cheap optical system which is effective in difficult conditions. Open sights will continue to be used mainly for fall back conditions, but telescopic sights will become general issue on weapons and collimator sights will also be used. It is likely that further developments in micro-channel plate image intensifiers will satisfy the requirements for lightweight monocular and binocular night viewing systems particularly as their reliability has improved substantially.

Appendix to Chapter 2

LIGHT UNITS

PHYSICAL AND PSYCHOPHYSICAL UNITS

Light units (illumination and brightness) are designed to express a visual response to a given level of illumination and as such are not easily related to the physical units of joules and watts. The various physical units involved in the measurement of radiation and their psychophysical equivalents are listed below:

Serial	Physical Quantity		Psychophysical Quantity	
(a)	(b)		(c)	
1	Radiant Energy	J	Luminous Energy	talbot
2	Radiant Density	Jm^{-3}	Luminous Density	talbot m^{-3}
3	Radiant Flux	W	Luminous Flux	lumen
4	Radiant Exitance	Wm^{-2}	Luminous Exitance	lumen m^{-2}
5	Irradiance	Wm^{-2}	Illuminance	lumen m^{-2} (lux)
6	Radiant Intensity (watt per unit solid angle)	W $ster^{-1}$	Luminous Intensity	lumen $ster^{-1}$ (candela)
7	Radiance	W $ster^{-1}m^{-2}$	Luminance	lumen $ster^{-1}m^{-2}$ or candela m^{-2}

The primary standard of the psychophysical system is the luminance of the surface of a blackbody radiator at the freezing temperature of platinum (2043.5°K) which is defined to be 6×10^5 lumen $ster^{-1}m^{-2}$.

LUMINOUS EXITANCE AND ILLUMINANCE

Luminous exitance and illuminance are normally expressed in lumens m^{-2} (lux). An earlier British unit is the Foot-candle (lumen ft^{-2}) = 10.764 lux. Some typical values are:

Sun (midday at the equator)	10^5 lux
60 W electric bulb	50 lux at 1 m

LUMINANCE

Luminance is the same as the more traditional term 'brightness'. Since 1 lumen $ster^{-1}$ = 1 candle, the unit of luminance is more commonly expressed in candela per square metre (candela m^{-2}).

Lambert's cosine law states that the amount of light scattered or received in a given direction falls off with the cosine of the angle between the normal to the surface and the given direction. A surface that scatters light in this way or one that emits light appears uniformly bright in all directions. This means that its luminance is the same in all directions. Such surfaces are known as uniform diffusers or Lambertian surfaces. It can be shown that such a surface whose luminance is 1 candela m^{-2} emits π lumens per $metre^2$. In practice many surfaces approximate closely to this type of behaviour. (A typical TV screen exhibits luminance variations ranging from 0.1 to 10^3 candela m^{-2} and the total light output is about 0.5 lumen).

Two other units for luminance are still currently used. They are:

Lambert, L. The Lambert is defined as the luminance of a uniform diffuser scattering or emitting 1 lumen per cm^2. Thus 1 mill-lambert (mL) is the luminance of a surface emitting 10 lumens per $metre^2$. Hence:

$$1 \text{ mL} = \frac{10}{\pi} = 3.18 \text{ candela m}^{-2}$$

Foot-Lambert (ftL). The foot Lambert is defined as the luminance of a uniform diffuser emitting 1 lumen per ft^2. Since

$$1 \text{ m}^2 = 10.764 \text{ ft}^2$$

$$1 \text{ lumen per ft}^2 = 10.764 \text{ lumens per m}^2 = 1 \text{ mL}$$

Note. A diffuse surface of reflectivity R illuminated with E lumens per unit area will scatter diffusely RE lumens per unit area. By definition this corresponds to a luminance of RE lamberts per unit area or RE / π candela per unit area. Thus for an incident illuminance of E lumens per m^2 ie E lux, the corresponding luminance of the surface is RE/10 mL or RE/π candela m^{-2}.

Trolland

The Trolland is a measure of retinal stimulation and is numerically equal to the product of luminance in candela m^{-2} and pupil diameter in mm. The relation between stimulation and luminance is shown in Fig. 2.16.

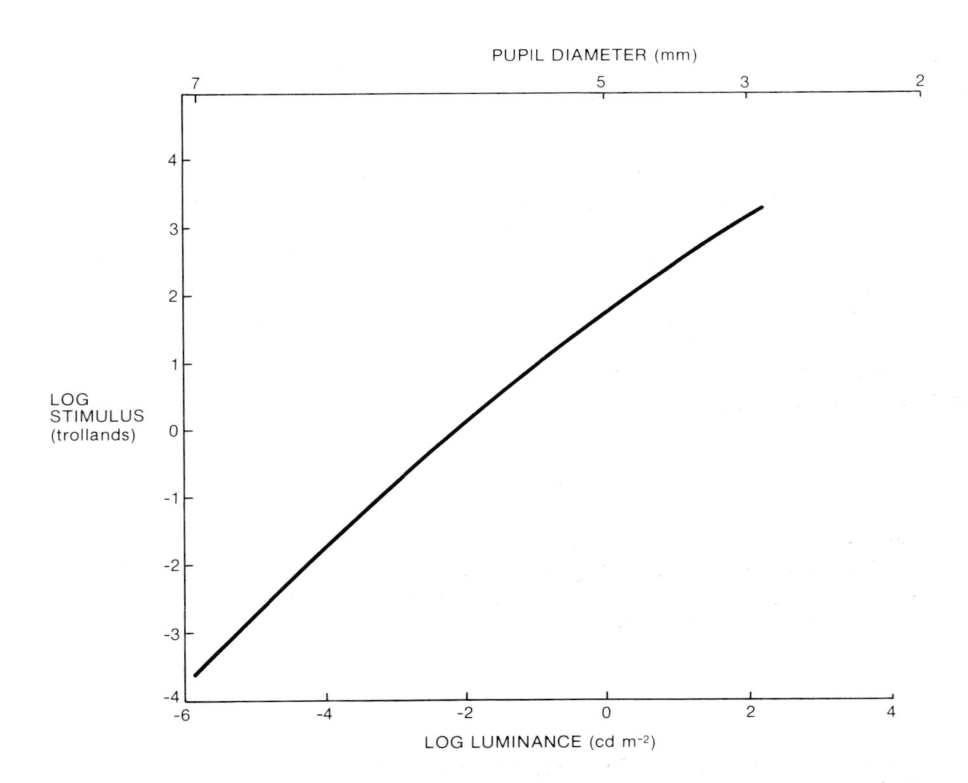

Fig. 2.16 Illuminance and eye stimulus

Equivalence of Physical and Psychophysical Units

The equivalence of physical and psychophysical units is a complicated question since it depends on a number of subjective variables: conditions of observation, age and experience of the observer, spectral content of the light. The numerical factor expressing the equivalence is wavelength dependent and at a wavelength of 0.555 μm it has the value 1 watt = 680 lumens. This means that a flux of 1 watt at 0.555 μm gives the same psychophysical sensation as 680 lumens. At other wavelengths in the visible region 1 watt produces a sensation less than 680 lumens.

SELF TEST QUESTIONS

QUESTION 1 What are the factors which determine the resolution of the eye?

Answer ....

....

....

....

....

....

QUESTION 2 If the limiting resolution of the eye is 60 cycles per degree, calculate the distance from a 19 inch, 625-line, 4 x 3 aspect TV screen at which the raster is just discernable.

Answer ....

....

....

....

....

....

QUESTION 3 Contrast variations on the image of a battle tank 2 m high viewed with the naked eye equate to 2 cycles. If the limiting resolution of the eye is 60 cycles per degree calculate the range of the tank.

Answer ....

....

....

....

....

....

QUESTION 4 How would the following surveillance systems perceive a battle tank 2 metres high which is 2 km away?

a. Unaided eye in bright sunlight?

b. Through a pair of hand-held 7 x 50 binoculars in conditions equivalent to snow in full moon?

Answer a. ....

....

....

....

....

....

b. ....

....

....

....

....

....

QUESTION 5 In a typical camera the F-numbers are 1.4-2-2.8-4-5.6-8. How is the exposure time changed in passing from one F-number to the next higher one to maintain constant illumination of the film? What other effects on performance would changing the F-number produce?

Answer ....

....

....

....

....

....

....

....

....

QUESTION 6 A tank is on the skyline in good daylight 5 km distant; it is 2.5 m high the visibility is good and the contrast reasonable. Estimate the optical magnification needed to:-

a. detect it
b. recognise it
c. identify it.

How would these calculated magnifications relate to hand-held instruments?

Assume eye resolution = 0.25 mils in the ambient light condition (1 cycle = 0.5 mils).

Answer ....

....

....

....

....

....

....

....

....

QUESTION 7 A sighting telescope has a magnification of x 3, a field of view of 8° and an object glass focal length of 100 mm. It is desired to double the magnification by substituting a new object glass and appropriate tube. What focal length would the replacement O.G. need to be and what would be the new field of view. Also what is the apparent field of view in both cases?

Answer ....

....

....

....

....

....

3.

Image Intensification

SEEING AT NIGHT

That the human eye cannot see in the dark is perhaps too extreme a statement to make as the previous chapter showed; but with the need to fight a 24 hour battle the ability to see well in the dark is very important. We have already dealt with the eye and the way in which it performs in some detail. However, as a starting point to this chapter it is necessary to restate the way in which the eye performs at low light levels.

The ability of the eye to discern fine detail - its resolution - is produced by the cones which are also responsible for colour vision. The range of response to luminance levels of the cones, known as the photopic response, is from 3 to 3×10^5 candela m^{-2}. 3 candela m^{-2} corresponds to poor daylight. Below this level the ability of the eye to discern detail falls off quickly. To protect the retina under very bright conditions the iris contracts limiting the amount of luminous energy falling on the retina. The upper limit of visual tolerance occurs at about 3×10^5 candela m^{-2}.

The rods are much more sensitive than the cones but because many rods are connected to an individual nerve fibre they are unable to discern fine detail. On the other hand this interconnection of the rods allows the signal to be integrated over larger areas of the retina thus improving the signal to noise ratio at the expense of the ability to discern fine detail. The rods operate at levels below 3×10^{-2} candela m^{-2}, the eye being fully dark adapted at about 3×10^{-5} candela m^{-2}. The time required for full dark adaptation is about 30 minutes and is the time taken for a build up of rhodopsin in the rods.

The problem then, can be very simply stated: it is to increase the illuminance level of the scene to a level at which the cones begin to operate and pick out enough detail to perform the required task. Table 1 shows the light levels in Lux and Foot Candles for various day and night light levels.

TABLE 1 Illuminance Levels by Day and Night in Lux and Foot Candles

Level	Lux	Foot Candles
Clear Sunlight	10^5	10^4
Full Summer Day (noon)	7×10^4	7×10^3
Cloudy Summer/Bright Winter Day (noon)	2×10^4	2×10^3
Heavily Overcast Winter Day (noon)	5×10^3	5×10^2
Good Interior Working Illumination	10^3	10^2
Winter Sunset (clear sky)	5×10^2	5×10^1
Twilight (dusk)	10^1	10^0
Deep Twilight	10^0	10^{-1}
Clear Moonlight (moon at zenith)	3.1×10^{-1}	3.1×10^{-2}
Moonlight (full moon)	10^{-1}	10^{-2}
Moonlight (quarter moon)	10^{-2}	10^{-3}
Clear Starlight	10^{-3}	10^{-4}
Overcast Starlight/Overcast Dull Rainy Night	10^{-4}	10^{-5}
Very Overcast Starlight	10^{-5}	10^{-6}

From Table 1 it can be seen for example that to enhance the brilliance of the scene from overcast starlight to twilight requires a gain of about 10^5. Some enhancement in vision at night can be gained by the use of purely optical aids such as night binoculars. However, this improvement is limited and although it may be adequate in moonlight it is not adequate in starlight and ultimately the aid makes things worse as the visual threshold is approached because of light losses in the optics. To obtain a very considerable improvement in vision at night some type of electro-optical device is necessary to give an image that is many times brighter than the scene when viewed with the unaided eye. There are two types of passive electro-optical surveillance device. One uses the ambient visual light plus some of the night sky radiation reflected from the scene and is described in this chapter. The other uses the thermal radiation emitted by objects in the scene and will be described in the following chapter.

NIGHT SKY RADIATION AND THE NEAR INFRA-RED

The radiation falling on a scene at night comes from a variety of sources. The moon which is reflected sunlight, planets, stars and sky glow all contribute.

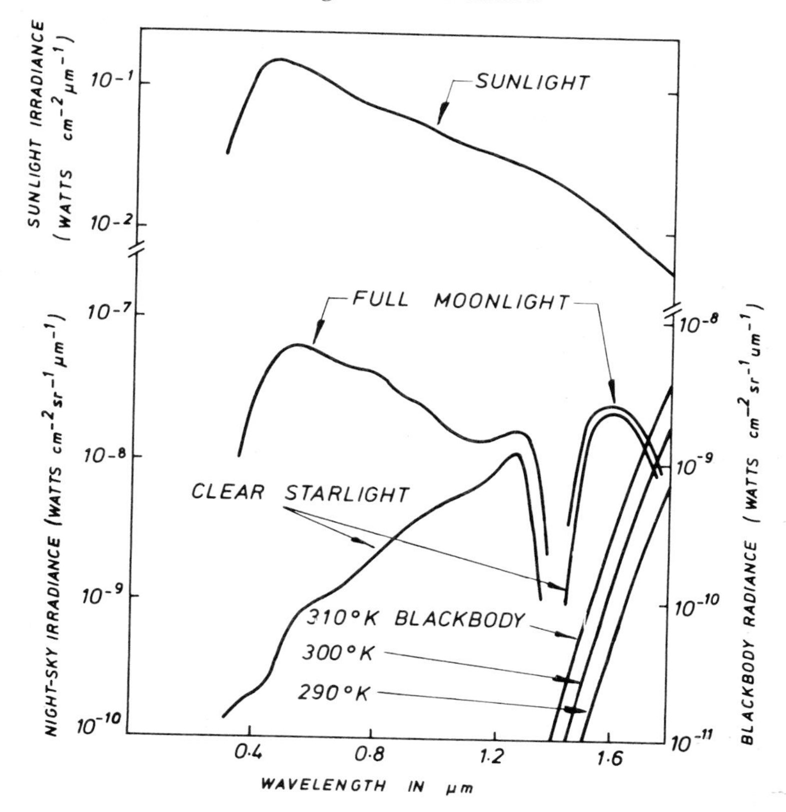

Fig. 3.1 Spectral distribution characteristics of sunlight, moonlight, starlight and some blackbodies

From the curves in Fig. 3.1 it can be seen that this radiation is present both in the visual part of the spectrum from 0.4 μm to 0.75 μm and also in the near infra-red part of the spectrum out to about 2.0 μm. Thus a night vision electro-optical device will be more effective if it can utilise the significant amounts of ambient near infra-red illumination. Such an electro-optic device must do three things. First it must collect photons at both visual and near infra-red wavelengths and convert them into electrons. Then it must increase the energy of the electrons by accelerating them through an electric field - the ampliciation process and finally it must convert the higher energy electrons back into photons at visual wavelengths.

A device which works in this manner is called an image intensifier and is shown in diagramatic form in Fig. 3.2.

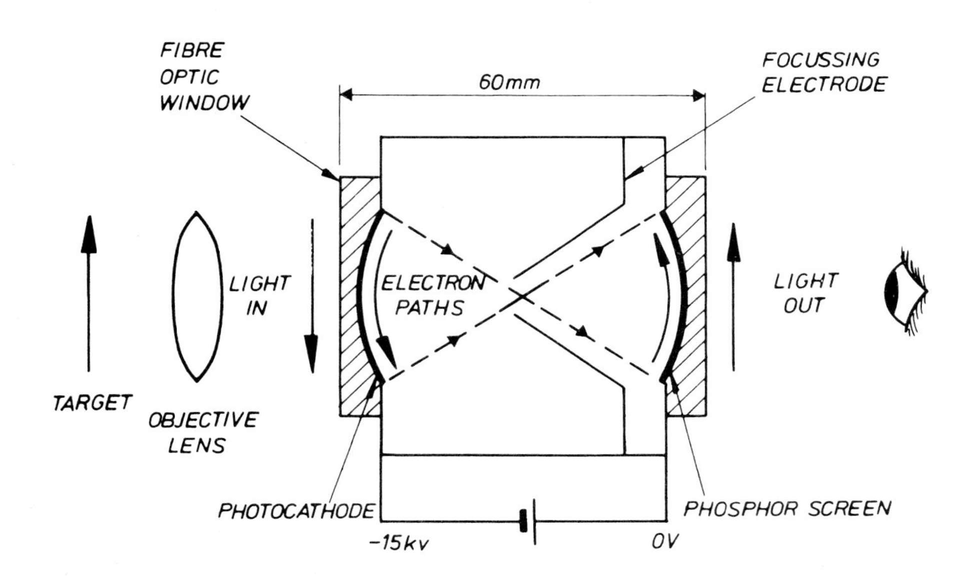

Fig. 3.2 Image intensifier

PHOTOCATHODES

Radiation reflected by the target is focused by an objective lens on to a photocathode. The photocathode is made from a substance which releases electrons when it absorbs electro magnetic radiation. This process is known as photo emission.

Photo Emission

There are three important facts concerning photo emission. It should be realised that no emission occurs if the frequency of the radiation is below a certain threshold value no matter what the intensity of the radiation. Then it is important to remember that the energy of the emitted electron depends only on the frequency of the radiation. Finally the number of electrons emitted per second (the current) is proportional to the intensity of the radiation.

The energy of the emitted electron is given by the Einstein formula

$$E = hf - w$$

Where E = energy of the electron in electron volts
f = the frequency of the incident radiation
h = Planck's constant (4.14×10^{-15} electron volt second)
w = Work function of the photocathode in electron volts

From this it can be seen that the minimum frequency (f_o) to cause photo emission occurs when $E = O = hf_o - W$

or when $hf_o = W$

W is called the work function and for metals is relatively high and f_o lies in the ultra-violet portion of the spectrum. To get photocathodes which work into the infra-red part of the spectrum requires materials with much lower work functions than simple metals. In practice photocathodes are made up of layers of several elements which act as semiconducting layers of low work function. It is of great practical significance that the number of electrons emitted from a given area per second is linearly related to the intensity of the incident illumination over a wide range since this allows good half tone reproduction in an imaging device.

Materials

A number of different combinations of materials have been used. An early cathode called S1 used silver, oxygen and caesium. It was the first successful material and was widely used in active infra-red systems. However it is rather insensitive and subject to thermal noise. The S25 photocathode consists of sodium, potassium, caesium and antimony. It has a much better response than S1 and is used in many current systems. A more recently developed material, caesiated gallium arsenide has a large and uniform wavelength response. It also has the advantage that it can be manufactured in a predictable manner using semiconductor processing techniques. The response curves of these photo emitters is shown in Fig. 3. 3.

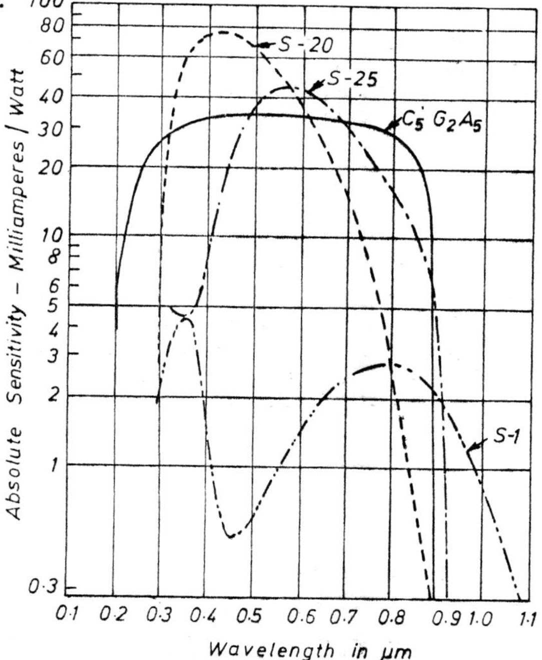

Fig. 3. 3 Photocathode performance

IMAGE INTENSIFIERS

Single Stage Tubes

In a single stage image intensifier the scene to be viewed is focused on to a photocathode. The electrons released are accelerated by an electric field of about 15 kv and directed onto a phosphor screen of cadmium activated zinc sulphide. The high energy electrons excite the cadmium atoms which radiate their excess energy as fluorescence in the visible portion of the spectrum. Thus at the phosphor there is a point by point correspondence with the infra-red image at the cathode. The initial image in the infra-red has now been converted into a visible image. The intensification of the image has come about because the acceleration of the electrons through 15 kv gives them enough energy for each electron to activate many cadmium atoms. For each photon of infra-red energy at the input there are many photons of visible energy at the output.

Characteristics

A good image intensifier should initially have a good conversion index. This is the ratio of the intensity of the visible image on the fluorescent screen to the focused image of the scene on the photocathode. It is the gain of the intensifier and a typical value for a modern tube is 50. In addition it requires clear resolution. This expresses the smallest detail which can be observed in the image. Resolutions of up to 40 line pairs per mm at about 20% contrast are possible, ie at 40 line pairs per mm the modulation transfer function is 20%. Resolution is greatly affected by the thickness of the phosphor screen. Then it needs a low background brightness. This is a measure of the brightness of the fluorescent screen when the photocathode is not illuminated. This should be as low as possible as it degrades contrast. It is caused by internally generated noise in the tube.

Cascade Tubes

The single stage tube suffers the disadvantage that the gain is insufficient for high speed detection and recognition of objects in a starlight scene and that the observer must be fully dark adapted for optimum gain. For this reason multi-stage or cascade tubes have been developed with which very high gains are possible. For example, to enhance the brilliance of a scene from overcast starlight to twilight requires an amplification of 10^5. This can be achieved with a 3 stage tube in which the single stage gain (conversion index) is of the order of 50. The individual stages are coupled together by bundles of fibre optics which constitute fibre optic face plates (Fig. 3.4). This allows the image to be transferred with minimum degradation from one curved surface to another of opposite curvature. The techniques for doing this are advanced and expensive. The culminative effect of internal noise which causes screen brightness imposes a practical upper limit of 3 tubes in cascade. Compared with single stage tubes a cascade tube suffers from lower picture quality because of the cumulative effect of the aberrations in the 3 stages and the inevitable increase in weight, size and cost. The combined effect of the 6 curved fibre optic face plates causes a marked decrease in brightness towards the edges of the screen.

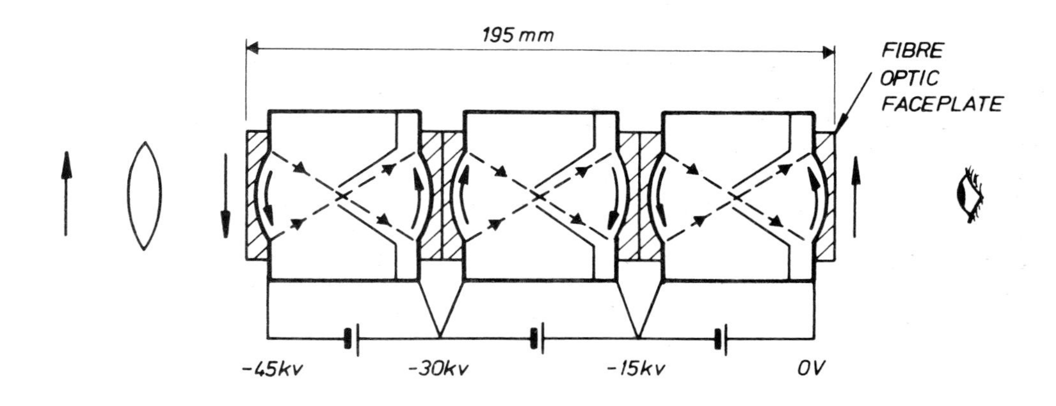

Fig. 3.4 Three stage (cascade) tube

Although the voltage shown in Fig. 3.4 is high (45 kv) the current drain of image intensifier tubes is low and a typical 3 stage tube will operate for over 60 hours from a U-2 sized battery.

Channel Tubes

A more recent method of obtaining high brightness gain employs the principle of secondary emission channel electron multiplication. When fast moving electrons pass through matter they collide with electrons in the outer orbit of atoms leaving a trail of ionisation in their wake. If a high voltage is applied across the material these secondary electrons will themselves be accelerated and cause further secondary emission. This phenomenon can be induced in tubes of semi-conductor glass or in tubes lined with semi-conductor glass (Fig. 3.5). These tubes are made up into fibre optic mosaics and inserted between a photocathode and phosphor screen (Fig. 3.6). Because the electrons are contained within the glass tubes the need for electrostatic focusing is eliminated. The electron gain depends on the ratio of the length to diameter of the channel rather than its absolute dimensions, the potential applied and the secondary emission coefficient of the material. Channel tubes are very small having typically a length of 500 μm and a diameter of 10 μm. Gains of up to 10^5 are possible from a single stage. The principle features of the channel plate image intensifier are a high brightness and small size and weight.

Channel tubes suffer from a number of limitations. First there is a loss of photo electrons in the glass without producing an output pulse. Then the secondary emission process is somewhat variable which results in an increased noise level. Occasionally an electron may transfer from one channel tube to another generating cross talk: this effect is not serious.

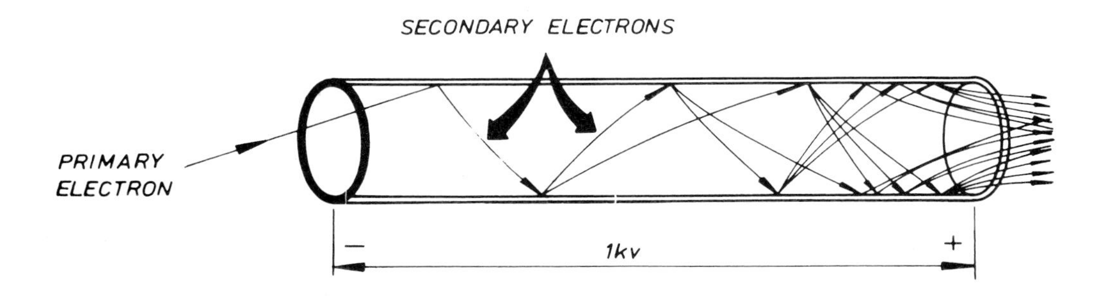

Fig. 3.5 Channel electron multiplication

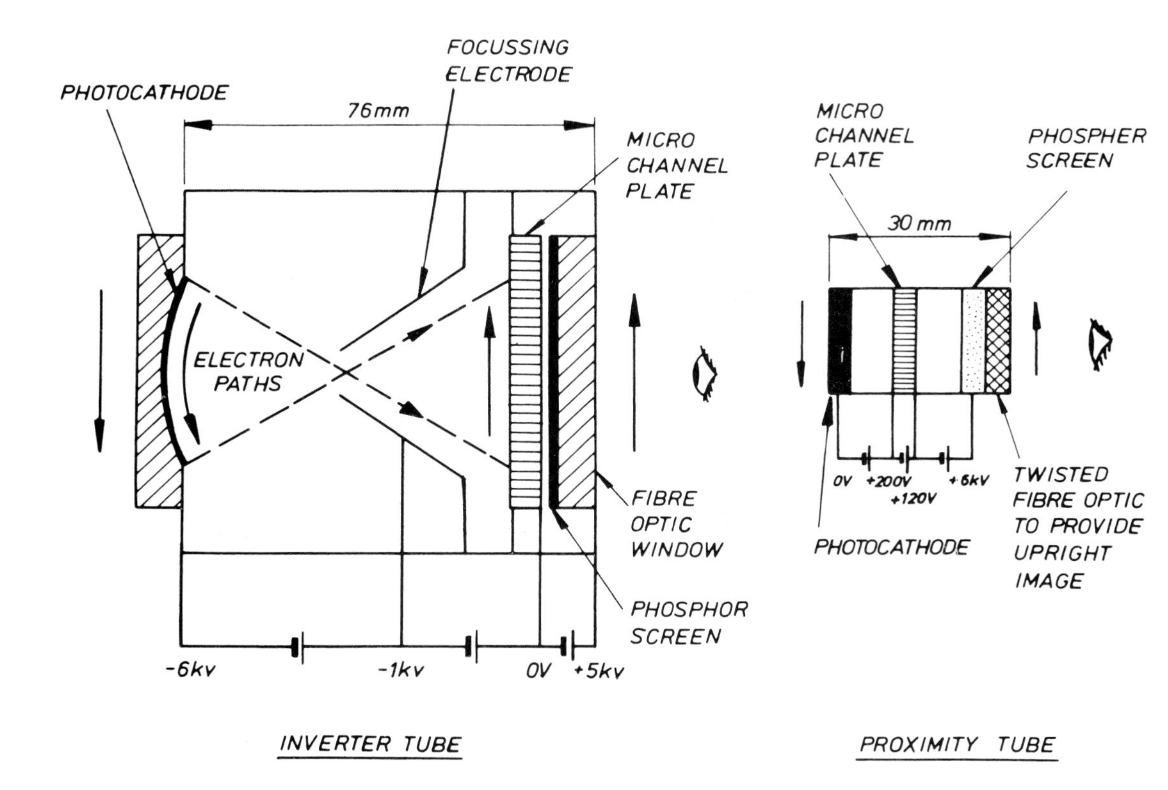

Fig. 3.6 Channel image intensifiers

Compared with the cascade tube the channel tube is lighter, smaller and superior in resolution. It does not suffer from saturation and whiteout due to over exposure like the cascade tube. Channel tubes on the other hand are twice as noisy at low light levels. The channel tube is now in use in military equipment where small size is a vital requirement, such as night flying goggles for helicopter pilots. A channel tube built to military specifications is liable to cost considerably more than a cascade tube with a similar performance.

Performance

The final form of the instrument will depend on the stated requirement. A conflict clearly arises, as in any optical viewing device, between field of view and resolution. Field of view is inversely proportional to the objective lens focal length; resolution is proportional to the objective lens focal length. The performance of an image intensifier has to be clearly stated and, since the device depends on ambient illumination, the light level must form part of the statement. Thus an imager will be required, for example, to recognise a MBT at a given range under clear starlight conditions. This is clearly shown in Fig. 3.7 which shows the UK Night Observation Device Type A (NOD A) and the Individual Weapon Sight (IWS).

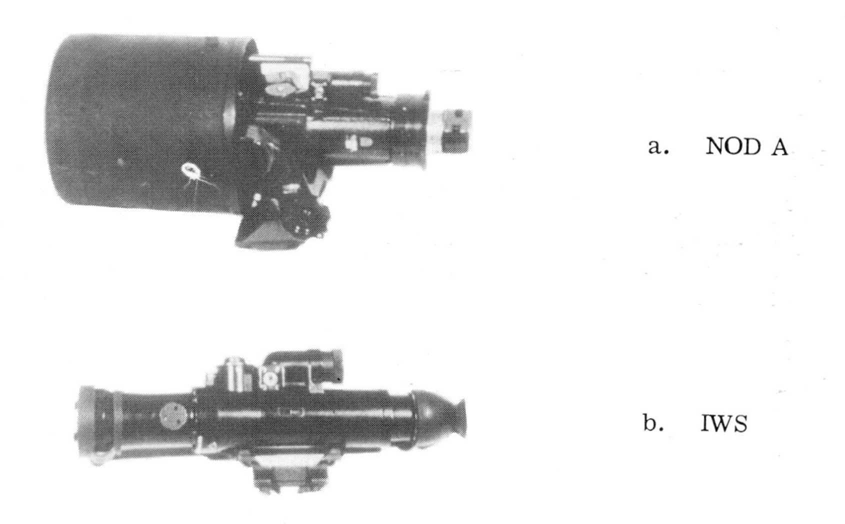

Fig. 3.7 UK NOD A and IWS

The only significant difference between the two is the focal length of the objective lens; both devices use similar 3 stage cascade tubes. NOD A with its large objective lens and resulting bulk and weight has a range four to five times greater than the IWS. The performance of an image intensifier can be improved by increasing the illumination by the use of searchlights, pyrotechnics or laser illuminators.

Development

Development of image intensifiers continues with the emphasis on the improvement of the photocathodes. The object of the effort being to push the response of the photo emitter further into the near infra-red in order to make use of all the available radiation. An improved photocathode gives the designer of an image intensifier two basic options; he can either keep the size of the objective lens, and thus the instrument, the same and get an improvement in range or he can accept no range improvement and achieve a reduction in size of the instrument.

LOW LIGHT TELEVISION

The development of television for outside broadcasting and film reproduction has led to the design of television camera tubes suitable for low light conditions. Basically, the television system uses the photons from the scene viewed and focuses them by means of an optical system onto a target from which they eject electrons. The effect of this is to leave behind in the material a positive charge for each electron emitted. Thus, there is left on the target a stored electrostatic image of the scene which corresponds to the intensity variations of the optical image. This charge image is then read out by an electron scanning beam to obtain the corresponding video current. Each positive charge draws an electron from the scanning beam: the greater the number of positive charges to be neutralised the more electrons are drawn and thus the higher the video current. Thus the video current varies with the brightness of the scene as the electron beam scans across it. The scanning process generates additional noise which may swamp the actual picture signal if the signal is not sufficiently large and can only be effectively avoided if a stage of image intensification precedes the camera tube. (See Fig. 3.8).

Low light television has the advantage that it is possible to provide remote viewing and multiple readout. Also video signals can be processed to provide contrast enhancement. Further, the television presentation makes it much easier to adjust tube brightness and amplifier gain in order to improve picture quality.

These advantages must be weighed against a number of disadvantages. Low light television is heavier and bulkier and more expensive than an equivalent image intensifier and also requires more electric power. The swamping of the video signal at low light levels by noise has already been mentioned. Another problem is the performance against moving targets. A system with good static resolution may have a slow response resulting in picture lag and blurring of the image.

Low Light Television Tubes

In the development of conventional television technology two main types of camera tube have evolved. The vidicon is used for industrial and military purposes and operates on the photo-conductive effect; the image orthicon is the main broadcasting type of camera and operates on the photo emissive principle. Both are storage type tubes and their action utilises the electric charges generated by the incident light during the relatively long intervals between successive scans of the image. Both types of camera tube are essentially daylight devices in that they operate from full daylight down to twilight conditions. It is, however, possible to produce variants of the basic types which operate down to much lower light levels.

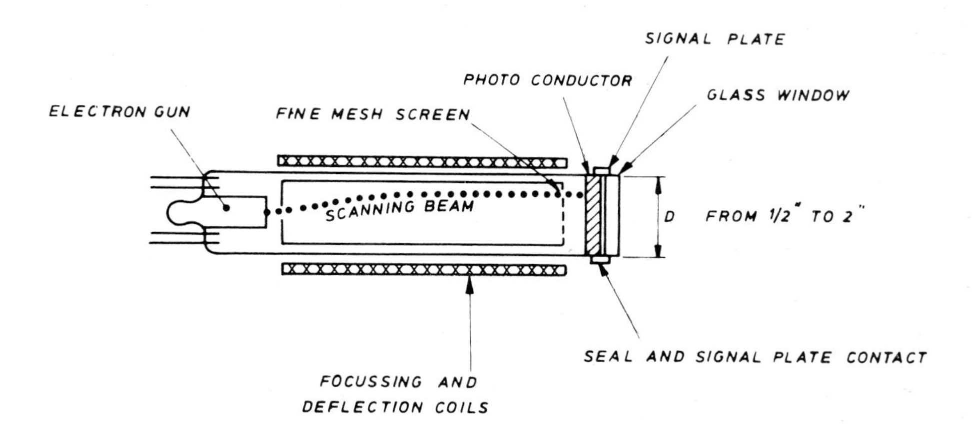

Fig. 3.8 Vidicon television camera tube

Vidicons

The low light response of the standard vidicon can be improved by introducing different photo-conducting layers on to the target. Of these the silicon diode array is the most sensitive. Sensitivity can be increased still further by the addition of image intensifiers in front of the camera tube but only at the expense of other characteristics such as resolution. An alternative approach is to build an image forming section into the camera tube itself. This is similar to a single stage image intensifier and gives rise to the following variants of the vidicon. The Secondary Electron Conduction vidicon (SEC vidicon) obtains its greater sensitivity by secondary electron production within the target material. In the Electron Bombardment vidicon (EB vidicon) very great sensitivity can be obtained through the large conductivity induced in the target by bombarding it with high energy photo electrons. The Ebicon is one version but the most sensitive is the Ebsicon which uses the silicon diode array already referred to. All of these variants can be used with image intensifier stages in front of them.

Orthicons and Isocons

The image orthicon is a larger and more complicated tube than the vidicon and gives high quality daylight performance. For low light work the image isocon has been developed from the orthicon. It provides excellent resolution and compares at low light performance to the SEC vidicon and Ebsicon. Image isocons can also be used with image intensifier stages in front of the camera tube.

Self Scanning Imaging Systems - Charge Coupled Devices

During the last decade a new type of silicon integrated circuit has been developed which has affected the design of television cameras for certain purposes by offering the possibility of smaller, much more rugged devices.

A charge coupled device (CCD) is, in effect, a high density information store with capacity of about 10^5 bits per sq cm and finds its main application in computor systems. The information is stored as electric charges under a linear or two-dimensional array of closely spaced electrodes. Subsequently the stored charges can be transferred to the output in sequence in shift register fashion by the application of clock voltage pulses.

Signals can be introduced into a CCD either electrically or optically. Thus CCDs may be used as solid state optical imagers which do not require scanning by an electron beam. This eliminates the need for the current, relatively long, highly evacuated glass television camera tubes and also the requirement for mains power.

The CCD is made from silicon and this restricts its use to the near infra-red region, down to about 1.1 m. However, the response of silicon at these wavelengths is superior to most photocathodes (vide the Ebsicon) so that CCD cameras are useful under low light conditions. Typical dimensions of one of these small cameras is of the order of 15 cm x 8 cm x 5 cm and a weight of just under a kilogram without the battery. The performance approaches that of a conventional camera.

Applications

It will be apparent from the forgoing that a wide range of camera tubes is available. The selection of a tube for a particular task will depend on precise characteristics required. Low light television has a wide range of applications. With the ability to control cameras remotely, low light television is ideal for static surveillance tasks. It has also been used in MBTs such as the Canadian Leopard. A modern application is in Hele-Tele which is used for aerial surveillance of crowds and traffic and it is also fitted in a number of military aircraft for night flying.

LASER ENHANCED VIEWING (LEV)

Principle

For specific applications at night or when there is little or no natural light, it will be necessary to illuminate a scene by artificial light in order to recognise and identify detail. Pyrotechnics or a remote searchlight can be used for this purpose. For other uses it will be more convenient to integrate viewer and illuminator. In this case a pulsed light source and a gated viewer are required so that the effects of back-scattered light from the atmosphere, which would otherwise

Fig. 3.9 Hele-Tele in stabilised mount

affect the image contrast, are reduced. In principle, the viewer remains switched off until the illuminating pulse has been reflected back from the target. The viewer is then switched on for a short period of time.

Laser Illuminators

A laser source offers the best solution for a pulsed illuminator. Ruby, gallium arsenide and neodymium types may be considered. Ruby (0.69 µm) is ideal in matching the photocathode response but it can only be used at low pulse repetition frequencies which would result in a marked 'flicker' in the viewer. Neodymium is a possible choice but its characteristic emission at 1.06 µm is outside the peak response of the image intensifier photocathode, and it will also require cooling. Thus the most common illuminator in this type of system is the gallium arsenide laser with a characteristic wavelength of 0.84 µm.

Viewers

Both low light television and image intensifiers have been used for LEV systems. The intensifier has the facility of being gated to accept only those electrons produced from the light reflected from the immediate vicinity of the target.

Gating

Various methods of gating the image intensifier are available. Standard tubes may be gated by pulsing between the photocathode and anode or in the case of a micro channel tube, across the micro channel plate. Special tubes exist with a gating grid close to the photocathode. This system allows a faster pulse rate and lower power consumption.

Operation

Once a target has been located it is necessary to adjust the delay between the operation of the illuminator and the operation of the viewer so that the viewer is switched on at the time when light from the illuminator, reflected from the target, reaches it. When this is achieved the target is said to be 'in the range gate', a term used in radar where the same technique is employed. If the target is within the range gate it will be fully illuminated and, depending on reflectivity, specific detail may be observed. However, if the background has a comparable reflectivity to the object to be viewed then the overall contrast will be low and the outline of the target may be lost. By adjusting the delay so that the target is just in front of the range gate, only energy reflected from the background will be accepted by the viewer and the object will be seen as a silhouette with a high contrast.

REFLECTIVITY

One of the advantages of using optical and electro optical devices for surveillance is that the image produced is very similar to that perceived by the unaided eye.

Thus the image is easy to interpret and there is no need for operator training. However, there is one effect which the use of the image intensifier has highlighted. The graph in Fig. 3.10 shows that there is a marked increase in the reflectivity of green vegetation in the near infra-red between 0.8 μm and 1.3 μm. This is caused by the chlorophyll which is present in all live green vegetation.

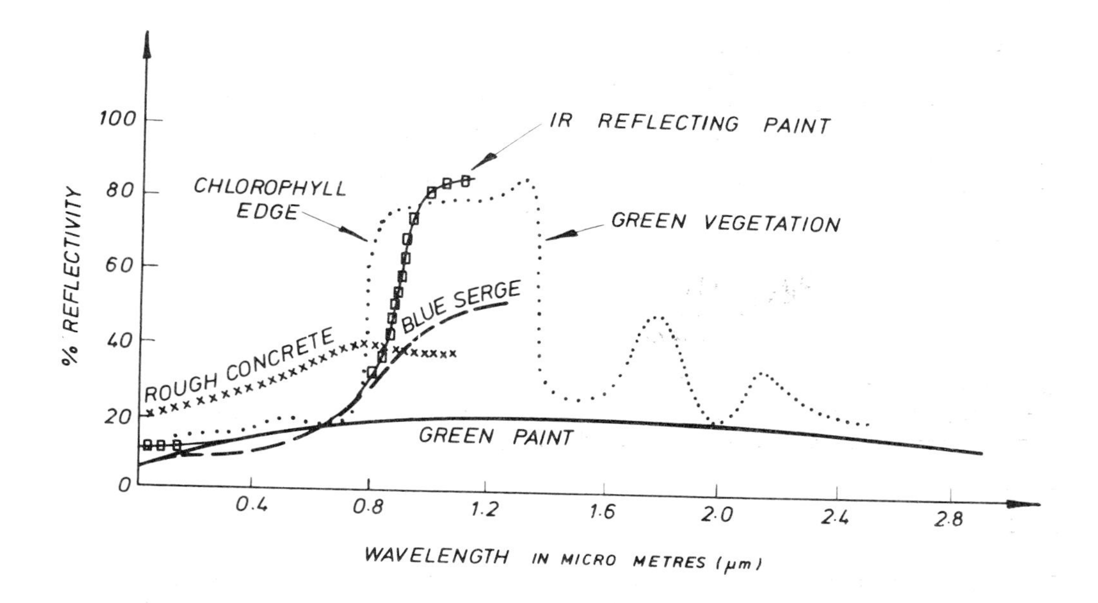

Fig. 3.10 Reflectivity of natural vegetation

Ordinary green paint for military vehicles was used because, at optical wavelengths, it had a similar reflectivity to vegetation. Thus vehicles could merge into the background. But when viewed through an image intensifier it was possible that the vehicle would stand out from the background because of the marked difference in reflectivity in the near infra-red. This has brought about the development of infra-red reflecting (IRR) paint, the reflectivity of which more closely matches that of vegetation at both optical and near infra-red wavelengths.

SUMMARY

Image intensification is a well developed technique for enhancing the performance of the eye at low light levels. Image intensifiers do not provide a complete answer to the all weather, day and night surveillance problem since they are as much affected by bad weather as is normal vision and they rely on some ambient light for their operation. Also, for long range surveillance they will be bulky. However, for short range work they are small and light and make ideal night sights for short range weapons and short range viewing aids. They are also considerably cheaper than thermal imagers. Thus, image intensifiers will continue to be found on the battlefield for short range work and where size and cost are important constraints.

SELF TEST QUESTIONS

QUESTION 1 What is purpose of an image intensifier?

Answer ....

....

....

....

QUESTION 2 What sort of gain can typically be expected from an image intensifier and why?

Answer ....

....

....

....

QUESTION 3 What are the principle sources of radiation utilised by an image intensifier?

Answer ....

....

QUESTION 4 Explain in simple terms the three principle functions of an image intensifier.

Answer ....

....

....

....

....

QUESTION 5 What is photo emission?

Answer ....

....

....

QUESTION 6 What are the three basic facts which govern photo emission?

Answer a. ....

....

....

b. ....

....

....

c. ....

....

....

QUESTION 7 What are the advantages and disadvantages of the channel plate tube compared with the cascade tube?

Answer - Advantages

....

....

....

- Disadvantages

....

....

QUESTION 8 What are the relative merits of lowlight television and image intensification?

Answer ....

....

....

....

....

....

QUESTION 9 Why has it been necessary to develop special infra-red reflecting paint for military vehicles?

Answer ....

....

....

....

....

....

QUESTION 10 What are the main roles for the image intensifier on the modern battlefield?

Answer ....

....

ANSWERS ON PAGE 190

4.

Thermal Imagers

THERMAL IMAGING

All bodies absorb and emit radiation over a continuous range of wavelengths which depends on the temperature of the body. This natural radiation, often called blackbody radiation, occurs mainly in the infra-red portion of the electro magnetic spectrum. The possibility of detecting objects through their blackbody radiation was considered before the 1939-45 war in both Germany and the UK. In 1938 both infra-red and radar technology were in their infancy and it was not certain which offered the better solution. Radar was chosen as the most promising by the UK whereas in Germany infra-red was adopted, particularly by the navy. As a result German infra-red technology was reasonably well advanced by the end of the war.

Since the end of the war there has been a considerable development in infra-red technology to meet military needs for missile guidance and for night vision devices to supplement image intensifiers.

Image intensifiers operate at wavelengths up to 1.2 μm and depend for their effectiveness on the level of ambient radiation. Their performance can be improved with the use of artificial illumination, either white light, infra-red illumination or laser enhanced viewing but such active systems reveal the observer's presence to an enemy also equipped with night vision devices. Interest has therefore been focused on passive viewing devices which make use of blackbody radiation emitted by the target.

BLACKBODY RADIATION

The physical relationships which describe the thermal radiation emitted by an ideal radiator or blackbody are rationalised by Stefan-Boltzmann's Law, Planck's Law and Wien's Displacement Law.

Stefan-Boltzmann's Law

Stefan-Boltzmann's Law states that the power radiated by unit area of a blackbody maintained at an absolute temperature T is given by

$$W = \sigma T^4$$

Where W = total radient emittance Wcm^{-2}

σ = Stefan-Boltzmann constant (5.67×10^{-12} Wcm^{-2})

T = absolute temperature ^{o}K

Planck's Law

Planck's Law gives the spectral composition of the radiation emitted by a blackbody at a given temperature. It is a complicated mathematical function and is best represented by the graph in Fig. 4.1.

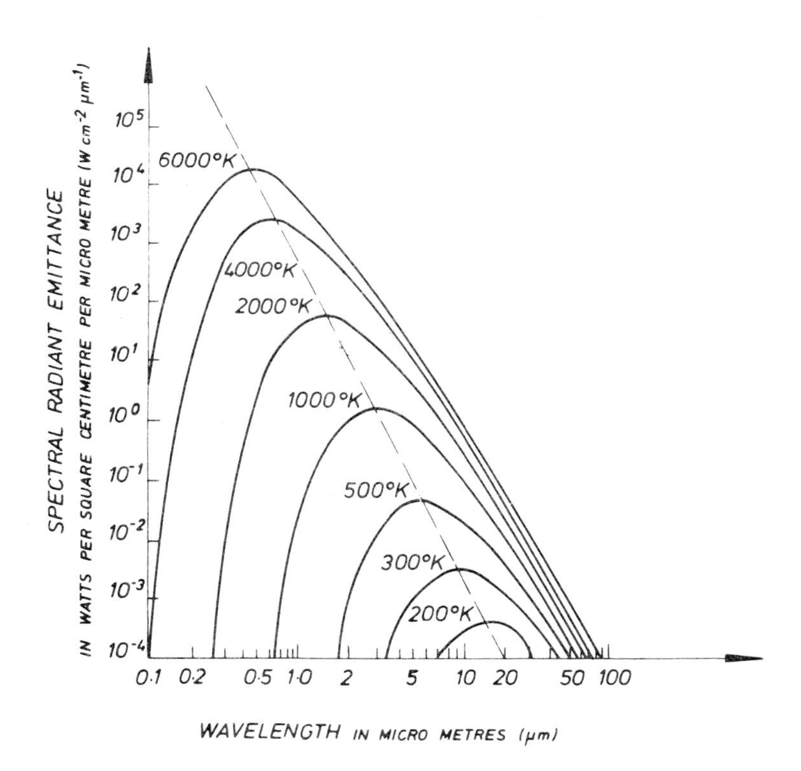

Fig. 4.1 Blackbody radiation

Wien's Displacement Law

Wien's Displacement Law relates the wavelength λm corresponding to the peak of the radiation curve in Fig. 4.1 to a given temperature T.

$$\lambda_m T = \text{constant} = 2897 \quad m\,^{o}K$$

Thus for a body at $300^{o}K$ ($27^{o}C$) $\lambda_m = 10$ m
and for a body at $3000^{o}K$ $\lambda_m = 1$ m

The graph at Fig. 4.1 reveals a number of important facts about blackbody radiation. The hotter a body is, the more energy it radiates in total and indeed at a given wavelength. A hotter body also radiates over a greater range of wavelengths and its radiation peak occurs at a shorter wavelength.

These mathematical relationships are for perfect radiators. In practice, most bodies do not radiate so effectively. Thus it is necessary to allow for the emissivity of the surface ε where ε is always < 1.

$$\text{Thus for a practical body } W = \varepsilon \sigma T^4$$

ε depends on the material, wavelength and to a lesser extent, temperature. Most metals for instance have low emissivities, say 0.4 at 1 μm decreasing to 0.04 at 10 μm. On the other hand, metals tend to be excellent reflectors of infrared whereas non metals tend to be poor reflectors particularly at long wavelengths.

The effectiveness of a surface as a radiator depends upon its temperature. In practice this temperature will be affected by a number of factors which include the incident radiation and neighbouring sources of heat, the absorptivity of the surface, its heat conductivity and thermal capacity of the material and then of course the size and shape and material of the target. Ambient conditions such as convective cooling or heating by wind, cooling by rain or condensation of dew and subsequent evaporation also have an important effect.

A surveillance system detects the presence of a target as a result of contrast or some discontinuity in the background. Contrast in the visible part of the spectrum occurs because of reflectivity variations in the scene. The analogy in the thermal part of the spectrum is not quite as simple since the thermal radiation coming from any point in a scene is the sum of the blackbody emission from that point and the reflection of ambient thermal radiation from neighbouring objects which also falls on the scene. Thus thermal contrast arises from the sum of variations in both emissivity ε and reflectivity r. However $\varepsilon + r = 1$ and thus variations in the two characteristics cancel out when all points in the scene have the same temperature. This is very rarely the case and thermal contrast arises from temperature variations in the scene. Whilst these variations may be small, the radiant emittance of two bodies at temperatures close to one another will differ considerably because radiant emittance is proportional to the fourth power of temperature.

Contrast is degraded with increasing range by unwanted radiation scattered into the same path as the wanted radiation and for large ranges contrast decreases exponentially with range. This is particularly the case for visible radiation, the visual range being defined as the range at which contrast falls to 2%, the lower limit for the eye. For infra-red radiation, scattering is much less, particularly at longer wavelengths so thermal imagers can, in general, outrange visual systems provided that the temperature variations in the scene are not too small. For satisfactory performance a thermal imager must be capable of detecting temperature differences of less than 0.5°C.

TRANSMISSION OF INFRA-RED RADIATION THROUGH THE ATMOSPHERE

For any transmission system which has to work over long ranges it is important to understand the way in which the transmission medium affects the radiation. In the case of infra-red surveillance systems the ranges may be several tens of kilometers (for aerial systems) and the transmission medium is the atmosphere. The two principle mechanisms which affect the transmission of infra-red radiation through the atmosphere and cause attenuation are scattering and absorption.

Scattering

Scattering is caused by the molecules of the medium, and in the case of the atmosphere, by the presence of particles of dust, smoke, haze, fog, rain and ice. The amount of scattering depends upon the wavelength of the radiation, particle size and concentration and on the refractive index of the particle. For IR transmission through the atmosphere in the wavelength range of 1-20 μm the most significant scattering is caused by particles whose size is of the order of the wavelength of the radiation. Thus fog, cloud, dust and rain where particle sizes range from about 1 to 50 μm are a problem. Where the wavelength of the radiation is significantly greater than the particle size there is relatively little scattering. Thus in a fine mist, where particle sizes are of the order of 2 or 3 μm and the visible range is greatly reduced, a thermal system working at 8-13 μm will be hardly affected. But if the mist persists and the water droplets increase in size to the order of 10 μm or greater, then the thermal range will fall and will not significantly exceed the visual range.

Absorption

The molecules of a medium absorb electro magnetic radiation at characteristic wavelengths. In the case of the atmosphere, the chief constituents which absorb strongly in the infra-red bands are carbon dioxide, water vapour and to a lesser extent, ozone. A typical transmission curve for the atmosphere is shown in Fig. 4.2 and it will be noted that there are certain bands of wavelengths where the atmosphere is very absorbent. There are also three important "windows" in the atmosphere - one which covers the visible portion of the spectrum and two in the infra-red approximately from 3 to 5 μm and from 8 to 13 μm. Clearly any thermal system must operate in one of these windows for efficient transmission.

PILKINGTON
GROUP

In the context of transmission it is vital to use components such as lenses which are transparent over the wavelength range used. It is worth noting that glass is no use at all since it is completely opaque to radiation of wavelengths greater than 2.7 μm. There are a number of materials which are transparent to infra-red radiation but many of them are unsuitable for use in military equipment. A commonly used material is germanium and the large objective lenses made of this material are very expensive.

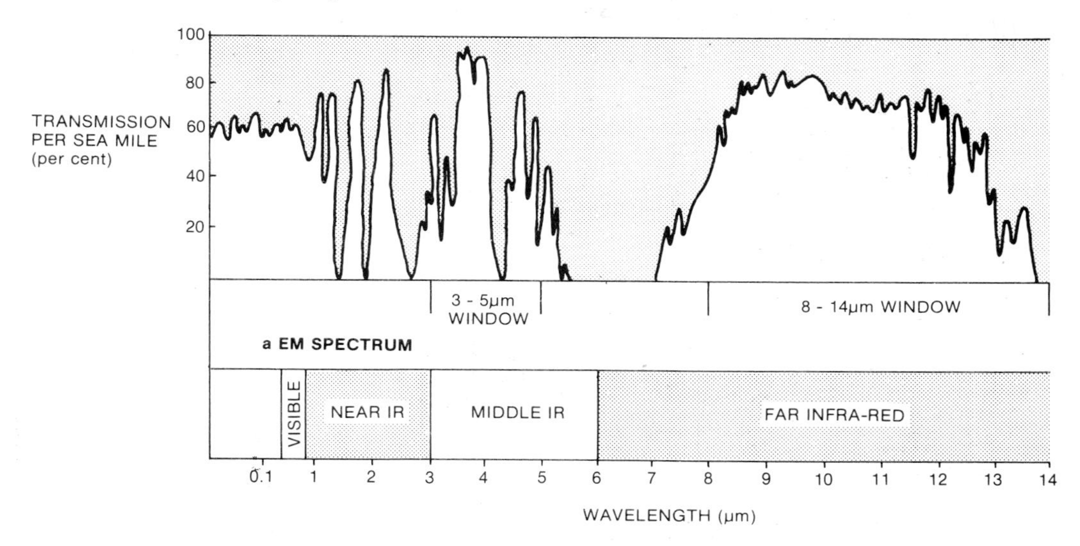

Fig. 4.2 Atmospheric windows

Infra-Red Detectors

The detector forms the principle element in any infra-red surveillance system and the rest of the equipment is built round it. The use of the photo emissive effect for detection of infra-red radiation at wavelengths greater than 1.2 μm is not possible because the energy of an infra-red photon at longer wavelengths is less than the lowest work function known at present. Other means have to be adopted and the two main types are thermal detectors and photon detectors.

Thermal Detectors

Thermal detectors respond to radiation by measuring the rate at which energy is absorbed. The subsequent temperature rise produces some physical change such as a thermal emf, a change of resistance or an increase in pressure of a small volume of gas. Their response is independent of the spectral content of the radiation; they respond equally well to all wavelengths. But, since the time taken to reach thermal equilibrium depends on the thermal capacity of the system their response is inevitably slow and they are unsuitable to detect moving targets. As a result they are seldom found in military systems with the exception of pyro-electric detectors which are now being successfully used in conjunction with vidicon tubes.

Photon Detectors

Many semi-conducting compounds suffer a reduction in electric resistance when they absorb photons whose energy exceeds a critical value. This is called photo conduction and is the most widely used method for detecting infra-red radiation. In all substances, the electrons in the molecules have discrete amounts of energy. In materials which are good conductors there are significant numbers of higher energy electrons which are only weakly attracted to the nuclei. These electrons are relatively free to move under the influence of an applied voltage and the material conducts electricity. In insulators there are no electrons available to act as current carriers. All the electrons are in the valance bands and high energies are needed to get the electrons out of the valance bands. In most semi-conductors the electron population of the conduction band is small and the resistance of the material is high. However, the difference in energy level between the conduction and valance bands (the band gap Eg) is not too large. This is represented graphically in Fig. 4. 3. To change the resistance of the material electrons must be raised from the valance band to the conduction band by supplying them with an amount of energy at least equal to that corresponding to the band gap. This may be accomplished in certain materials called photo conductors by exposing the material to radiation whose photons possess energy in excess of the band gap. A decrease in resistance will occur if $Eg > hf$. If Eg is expressed in electron volts, the critical wavelength below which photo conductivity can occur is given by

$$\lambda_c = \frac{1.241}{Eg} \quad \mu m$$

Hence the smaller Eg the larger the value of λ_c. Critical wavelengths for a number of semi-conducting materials are listed in Table 1.

TABLE 1 Critical Wavelengths (λ_c) below which Photo Conduction will occur for some Semi-conductors

Material	Eg (EV)	λ_c (μ m)
Silicon	1. 1	1. 1
Germanium	0. 7	1. 65
Lead Sulphide	0. 4	3. 1
Lead Selenide	0. 5	2. 5
Lead Telwede	0. 22	5. 6
Gallium Arsenide	1. 4	0. 89
Indium Antimonide	0. 2	6. 2

From the equation $\lambda_c = \frac{1.241}{Eg}$ µm it can be calculated that to detect radiation at 10 µm an energy gap of less than 0.12 eV is required and the development of suitable materials has taken considerable time. The most notable of these materials is cadmium mercury telluride (CMT); in appropriate proportions it responds over the range 6-12 µm.

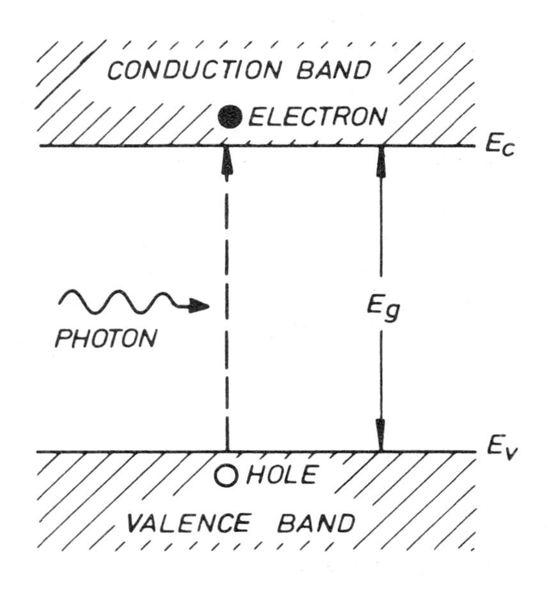

Fig. 4.3 Energy level for photo conduction

Noise in Photon Detectors

Noise imposes a limit on the detection capability of any system. The weakest signals which it is desired to detect must produce an output from the device sufficiently large for it to be distinguished from the noise. In photon detectors there are three sources of noise. Thermal noise is due to the random motion of the electrical charges in the detector: it is a function of the energy of the electrons and increases with temperature. Fluctuations in the number of electrical charges carrying the current is another source: it is most pronounced for semi-conductors with long wave response; this is because they have a low energy gap and it is possible for electrons to move from the valance to conduction band easily as a result of their thermal energy. The third source, photon noise, is not a function of the detector but arises because the photons of radiation from the scene arrive at the detector in a random manner which gives rise to noise: if all other sources of noise are removed it is photon noise which ultimately limits the performance of the system; cooling the detector reduces the noise generated within the detector; when the detector is cooled to about 77°K (the boiling point of liquid nitrogen) the photon noise limit is approached.

Performance of Photon Detectors

So that the performance of a photon detector could be stated it was necessary to have some means of quantifying performance. Originally Noise Equivalent Power (NEP) was used. This is the radiant power incident on the detector needed to produce an output equal to the noise output from the detector in the absence of a signal. A small NEP implies a good detector and vice versa.

A more common criterion is the detectivity D* (Dee-star). It is a measure of the ability of the detector amplifier system to produce a good signal to noise ratio from a standard infra-red signal. It is thus capable of being checked in any laboratory equipped with infra-red measuring apparatus. Since the output signal from the detector itself is usually very small the radiation from the infra-red source is interrupted or chopped by mechanical interruption. In this way an alternating signal is obtained from the detector which may readily be amplified. The size or sensitive area of the detector Ad, the frequency bandwidth of the amplifier Δf as well as the signal to noise ratio are all accounted for in the term for detectivity which is defined as:

$$D^* = \frac{Ad.\ \Delta f}{NEP} \qquad \text{Cm Hz}^{-\frac{1}{2}}\text{W}^{-1}$$

The larger the D* the better the detector. As the noise of the detector is reduced by cooling so the NEP falls which is reflected in an increase in D*.

The spectral response of photon detectors (in contrast with thermal detectors) increases with wavelength up to the critical or cut off wavelength λ_c when it falls abruptly to zero. This behaviour is explained by the fact that because the energy of an individual photon decreases with increasing wavelength the number of photons per unit energy must increase with increasing wavelength. The spectral response is related to D* and spectral response curves are plots of D* against wavelength. Figure 4.4 shows response curves for some of the more important detectors together with the photon noise limit for 290°K. A study of them shows that the D* for a given detector increases almost linearly with λ falling off abruptly near the cut off, increases as the detector is cooled and D* increases as the temperature is reduced virtually approaching the photon noise limit in some cases.

FAR INFRA-RED SYSTEMS

Electro magnetic radiation in the far infra-red portion of the spectrum (3-300 μm) is currently being exploited in the following applications of military interest.

Thermal Pointing

Thermal pointers were developed for use in conjunction with image intensifiers to improve target detection in conditions of poor visibility or through smoke and camouflage. There are no in service equipments as their development was overtaken by the progress made in thermal imaging. An early prototype consisted of

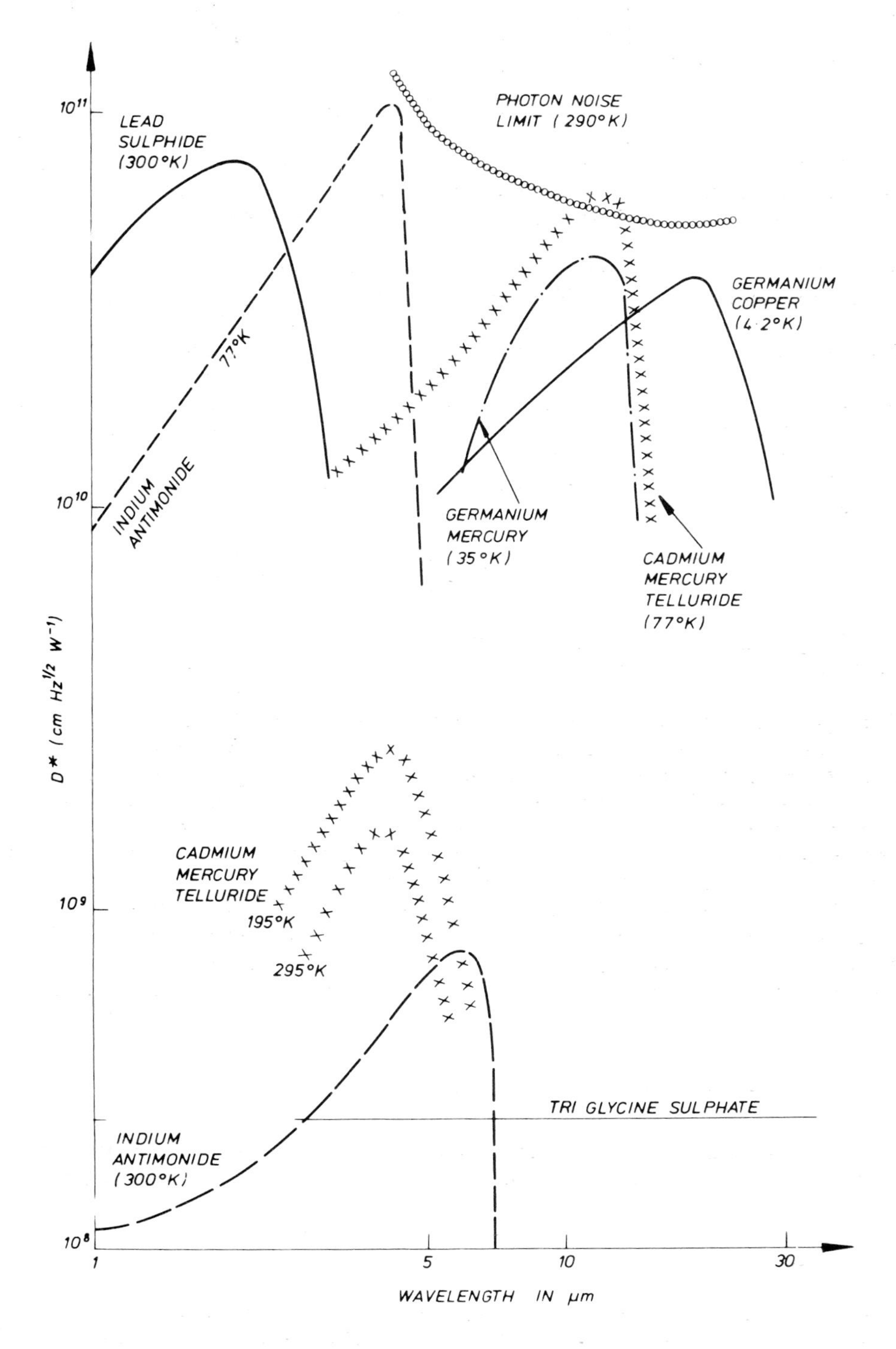

Fig. 4.4 Response of infra-red detectors

a single detector operating in the 3 to 5 μm band. It was scanned in the vertical plane only, the sight being slewed to give horizontal scan. When infra-red radiation from a possible target fell on the detector the output was used to switch on a light which was projected optically into the observer's field of view.

Remote Ground Sensors

Several types of intrusion alarm, including burglar alarms, use an active infra-red system. The alarm is activated when the beam of infra-red radiation is cut. It is also possible to use passive infra-red detectors in remote sensor systems.

Infra-Red Linescan (IRLS)

IRLS is an airborne system used in reconnaissance aircraft and drones to supplement conventional photography and give a night capability. The motion of the aerial vehicle provides the scan in one direction, mechanical scan in the form of a rotating prism is used to scan across the line of flight. The infra-red radiation is focused onto a detector and the output from this is used to drive a glow tube. The varying light from this tube exposes a film which is slowly drawn across it; when recovered and processed the film provides a permanent record for interpretation. IRLS is a low resolution system of the order of 1 m rad.

Missile Guidance

Infra-red homing or heat seeking missiles have been in service for some time now. A suitable detector, usually operating in the 3 to 5 μm band locks onto the hot exhaust of the target aircraft. The output signals of the detector are fed to control surfaces on the missile and used to guide it to its target. In some anti-tank guided missile systems the operator merely keeps his sight on the target while an infra-red system is used to track the missile. The tracker system evaluates the difference between line of sight to the target and to the missile and sends signals to the missile to bring it onto the line of sight to the target.

Thermal Imaging (TI)

TI is the application of far infra-red technology currently undergoing most development and it will therefore be covered in some detail in the paragraphs which follow. Two distinct trends can be discerned. The first is the use of cooled detectors of the cadmium mercury telluride (CMT) to give high resolution to the order of $\frac{1}{3}$ m rad and hence longer range imagery. The second is the use of pyroelectric detectors which give lower resolutions and shorter range imagery.

THERMAL IMAGING SYSTEMS

The two atmospheric windows that are relevant to thermal imaging are the 3 to 5 μm band and the 8 to 13 μm band. The choice of band depends on the characteristics of the target and the availability of detectors. Detectors were first developed which operated in the 3 to 5 μm band and it is only recently that CMT has been developed allowing the exploitation of the 8 to 13 μm band. From Fig. 4.1 it is clear that there is considerable radiation from hot bodies such as jet engines (1200°K) and vehicle exhausts (500°K) in the 3 to 5 μm band and systems have been developed to work in that band. However, in order to detect relatively cool targets such as men and cold vehicles, the 8 to 13 μm band must be used.

In Fig. 4.5 is a block diagram of a thermal imager. The lens system functions in exactly the same way as an optical lens system with the important difference that the lenses are made of a substance which transmits far infra-red radiation - probably germanium. The scanner and detector are the heart of the imager and will be described more fully below. The output of the detector will be a video signal and in many cases will be in the form of a standard television signal. The signal processing is therefore very similar to television signal processing and the display will be either a television monitor or for small applications some form of LED display.

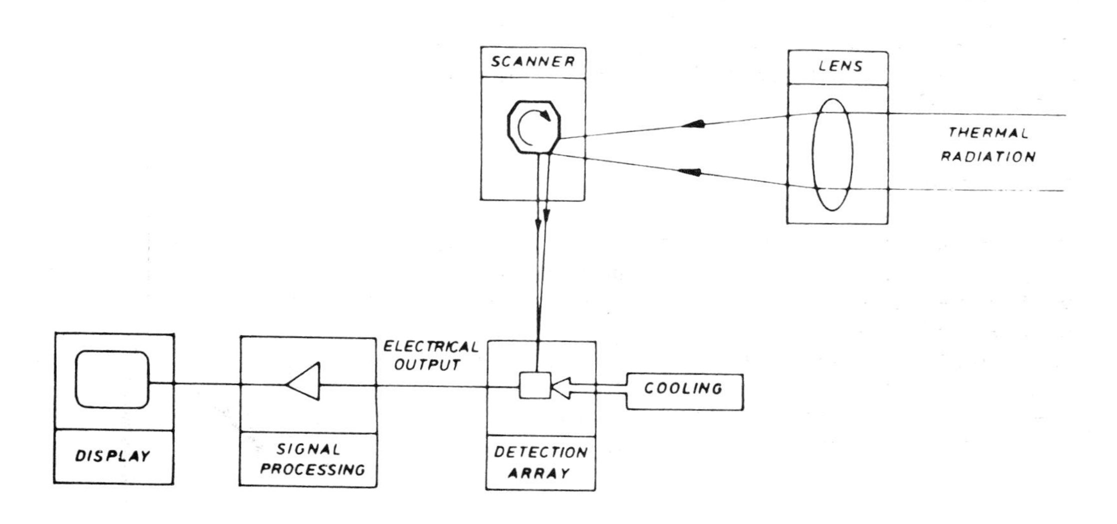

Fig. 4.5 Block schematic of thermal imager

Scanning Systems and Detector Arrays

The ideal detector would be one consisting of a mosaic of elements; this would resemble the eye and would not require scanning since the whole scene would be imaged on the array. A resolution of 0.5 m rad and a field of view of 20° would require about half a million elements (or picture points). Each detector would require its own pre-amplifier and each channel would require to have an identical response. The problem is that it is well beyond the capabilities of current technology to produce such an array with each channel having an identical response.

Going to the other extreme, a single element could be used and could be scanned across the whole scene. This system is shown in simplified form in Fig. 4.6. However, if the picture quality required is to be anything like TV an information bandwidth of about 5 MHz is required. CMT has one of the fastest response times of the photo conductors but its bandwidth is of the order of 200 kHz and is clearly inadequate. Early imagers built in the 1950's used single element detectors but their peformance was poor, particularly against moving targets.

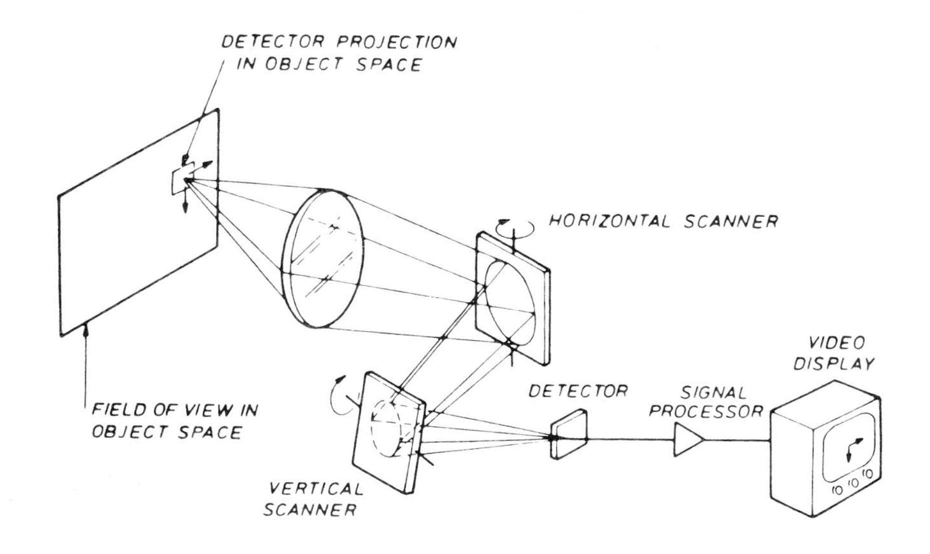

Fig. 4.6 Simplified single-detector, dual-axis scanner

The use of an array of detector elements is therefore essential at present if both high thermal sensitivity and picture quality are to be achieved. Here it should be noted that the use of a detector array with N elements improves the signal to noise ratio by $\sqrt{N}$. This is because the sampling time for any one picture is increased by a factor of N; the signal is increased N times and the incoherent noise by $\sqrt{N}$. The array may be arranged in one of two ways, as a parallel system or as a serial system.

The layout of the parallel array is shown at Fig. 4.7. It is mounted vertically and scanned horizontally. The whole scene may be scanned in one sweep provided the array contains enough elements for adequate resolution. The same result can be obtained by two interlaced sweeps and half the number of elements or with a series of contiguous horizontal sweeps covering the vertical field of view in a series of swathes from top to bottom as illustrated in Fig. 4.8. The output of each detector is amplified and displayed sequentially on a CRO or LED display. A typical cooled CMT detector would have 150 elements each 50 μm square. With a frame rate of 25 per second, a 300 line picture and suitable optics, such a system would have a sensitivity of less than 0.5°C and a range of 3 to 5 km.

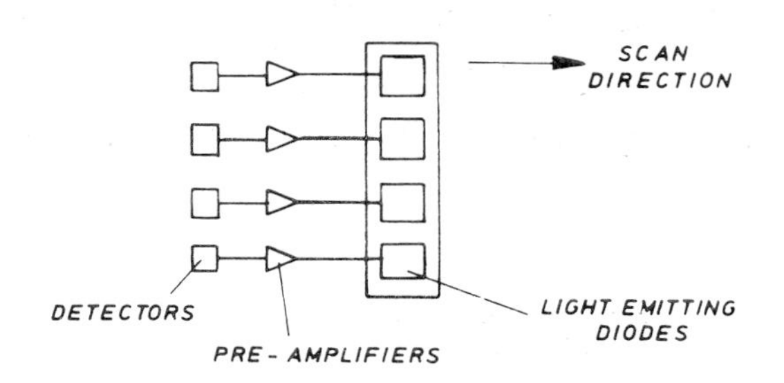

Fig. 4.7 Parallel scan processing

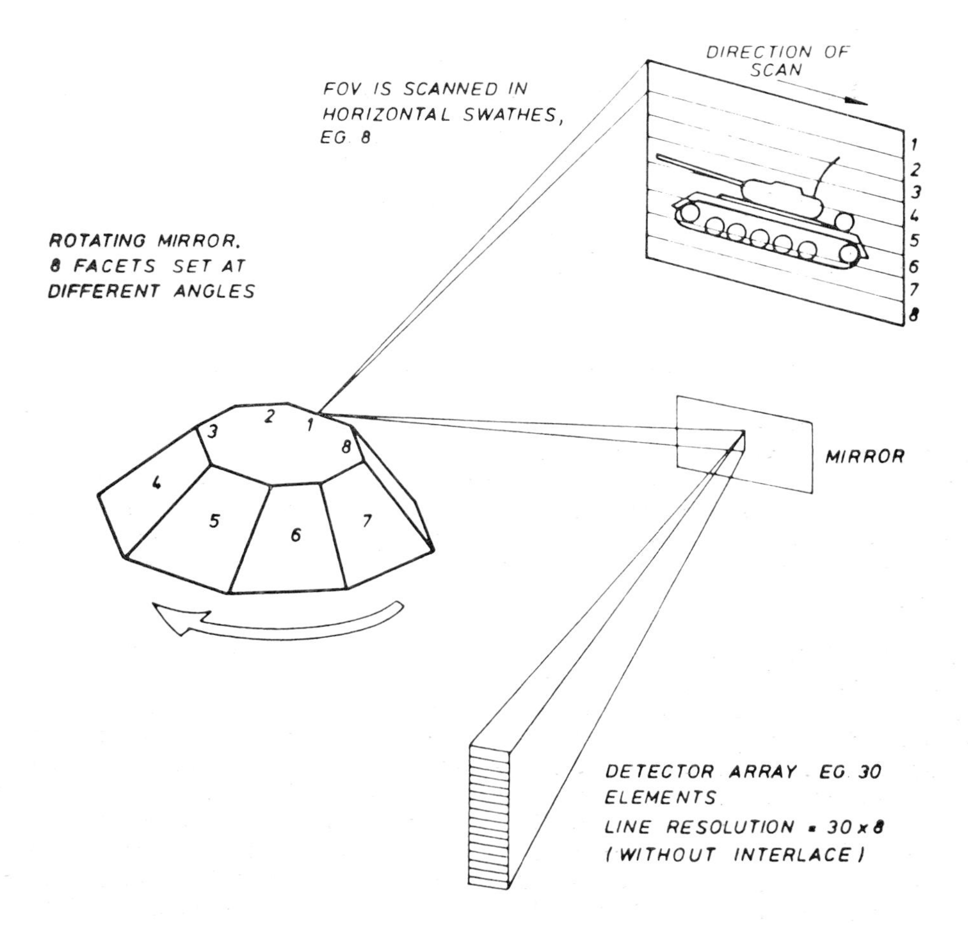

Fig. 4.8 Parallel scanning mechanism

The serial scanning array is mounted horizontally (Fig. 4.9) and each point of the picture is scanned by all the detector elements. The various outputs are approximately delayed and summed by an integrating delay line.

Each type of detector arrangement has its advantages and disadvantages. The parallel array is compact but variations in the responses of the individual elements introduces a type of spatial noise not present in the scene. The serial arrangement overcomes this defect by averaging out the detector variations; on the other hand the optical arrangements tend to be more complex.

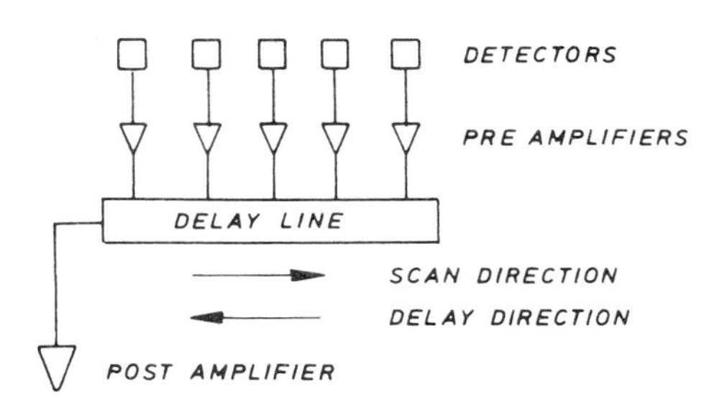

Fig. 4.9 Serial scan processing

Non-Scanning Systems

The complexity and delicacy of mechanical scanning systems suggests that alternative non mechanical methods would have been developed. The various attempts have resulted in too low a thermal sensitivity, resolution and response time compared with mechanical scanning systems. Television systems and image converter tubes, used at optical wavelengths, will not function satisfactorily at far infrared wavelengths without rather complex signal processing. This failure of performance is due to the relatively small thermal contrasts which require an almost impossibly uniform response over the surface of the detector. As for the mosaic array, the manufacture of detectors to this standard of tolerance is not feasible at present.

The Pyro-Electric Vidicon

The most successful non scanning system developed to date is the pyro-electric vidicon. This is a thermal detector and uses the pyro-electric effect where a change in material temperature produces a change in electric polarisation (or positive and negative charges on opposite surfaces). The material used, a single crystal of triglycine sulphate (TGS), is electrically polarised by the application of a dc field. The front face of the crystal is covered by a thin conducting layer which is held at constant voltage. Incident thermal radiation heats the material and produces polarisation changes in the crystal. Because the front face is held at a constant voltage the polarisation changes result in a varying pattern of positive charge over the rear face of the crystal. This is scanned by an electron beam as in the standard vidicon. The beam current as it scans across the crystal varies as the distribution of thermal radiation from the scene.

The detector response is largely independent of wavelength between 2 μm and 400 μm and temperature sensitivities of 0.2°C are now possible. Since the device is thermal the temperature of the sensing layer must be allowed to become uniform between frames if image smear is to be avoided. Some devices employ a mechanical shutter for this purpose, in others the surface is continuously rocked. Another problem is 'thermal spread' in the crystal caused by conduction from the warmer areas. This effect can be reduced by using a mosaic of TGS elements. Overall, the performance of the pyro-electric vidicon is inferior to a thermal imager using a cooled photo conductor; data rates are lower and spatial resolution is limited. A pyro-electric vidicon is unlikely to have a range greater than 1 km but for short range work its advantages are its lack of cooling and greatly reduced cost.

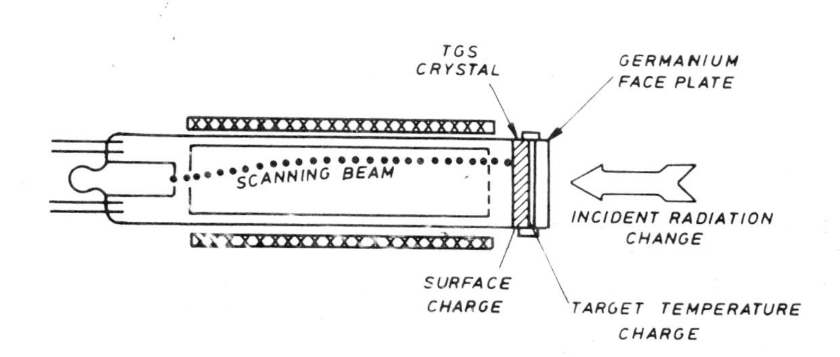

Fig. 4.10 Pyro-electric vidicon

Cooling

Thermal imagers working in the 3 to 5 μm band targeted against relatively hot objects need cooling to temperatures which can be attained by electrical methods. However, thermal imagers for battlefield use working in the 8 to 13 μm band require cooling to very low temperatures. This is achieved by immersing the back of the detector in a liquid gas which is allowed to boil to ensure a steady temperature.

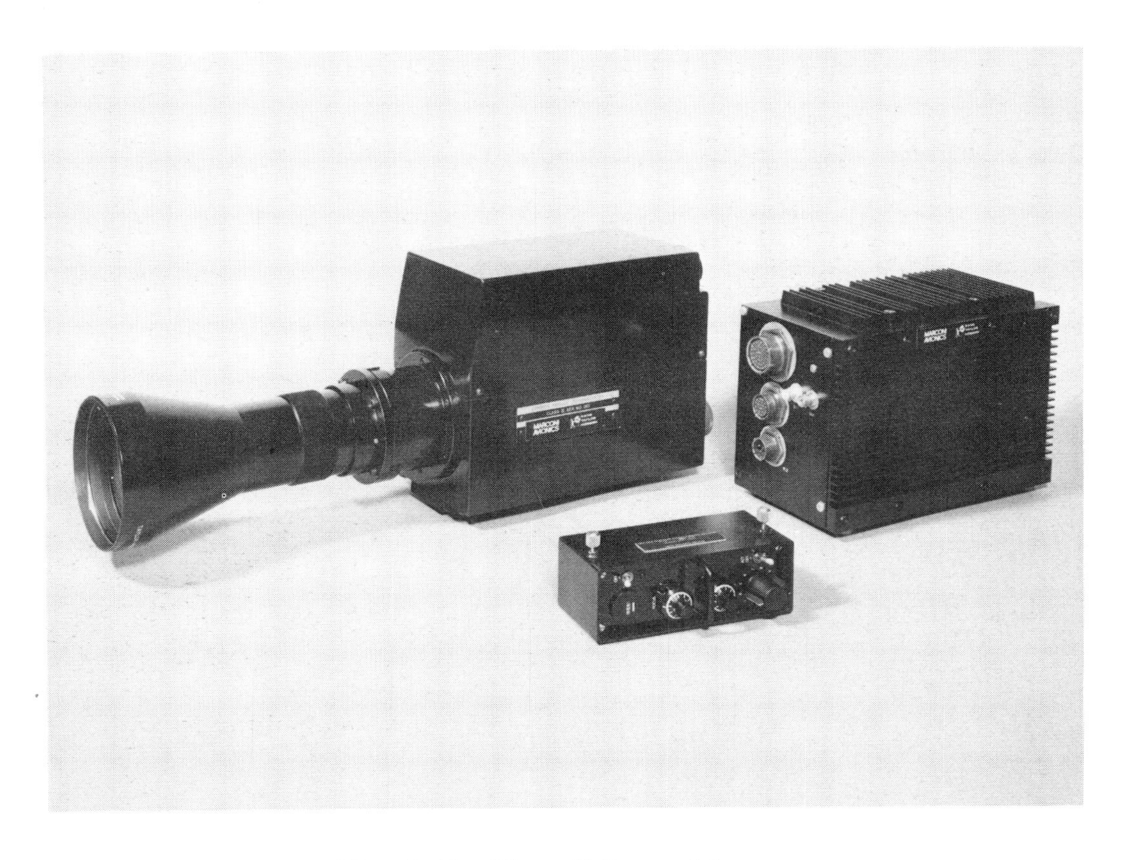

Fig. 4.11 Class II thermal imager

Liquid nitrogen can be obtained in bulk and transferred to the detector using the Leiden Frost liquid gas transfer process. A more practicable system for field use is to produce liquid nitrogen at the detector by attaching it to a mini-cooler. The mini-cooler is suppiled with compressed gas at high pressure from a cylinder and makes liquid gas by using the Joule-Thompson cooling-by-expansion phenomenon. Recharging the gas cylinders in the field from mobile compressors is relatively simple but imposes an additional logistic burden. It is possible that in the future, vehicle borne imagers will be supplied with liquid gas from a cooling engine operating on the Stirling or a similar cycle.

Fig. 4.12 Landrover viewed through a cooled thermal imager

SUMMARY

The introduction into service of thermal imagers provides the first 24 hour passive surveillance and target acquisition systems on the battlefield. Thermal imagers used for surveillance will be able to detect targets out to the normal limits of visibility and thermal imaging sights are being produced which will be able to acquire targets out to the maximum range of all direct fire weapons.

Although the thermal imager is a passive system which does not rely on ambient illumination it is not an all weather system. Under clear conditions thermal imaging will usually outrange visual surveillance and thermal imaging will penetrate haze, battlefield smoke and light mist. However, fog and rain, where the droplet sizes approach the far infra-red wavelengths cause severe scattering and reduce the thermal range to little better than the visual range. Thus the thermal imager does not provide the complete answer to the all weather problem but requires complementing by a system such as radar which is less effected by fog and rain.

Thermal imagers currently require quite large amounts of power, measured in tens of watts. This may be reduced as development proceeds. However, the requirement for cooling will remain as long as photo conductors are used, with its additional logistic burden. And finally thermal imagers are expensive, current equipments costing five to ten times as much as the image intensifiers they replace.

SELF TEST QUESTIONS

QUESTION 1 Express in simple terms the three basic facts of blackbody radiation.

Answer a. ....

....

b. ....

....

c. ....

....

QUESTION 2 What causes thermal contrast?

Answer ....

....

....

....

....

....

QUESTION 3 What are the two principle mechanisms which affect transmission through the atmosphere?

Answer ....

QUESTION 4 Does the atmosphere transmit all electro magnetic radiation?

Answer ....

....

....

....

QUESTION 5 Why is the photo emissive effect not used for far infra-red detection? What two principle mechanisms are used?

Answer ....

. . . .

. . . .

. . . .

. . . .

QUESTION 6 What is it which limits the detection capability of a photon detector and how are the best results achieved?

Answer . . . .

. . . .

. . . .

. . . .

. . . .

. . . .

QUESTION 7 What would be the arrangement of an ideal detector and why has one not yet been made?

Answer . . . .

. . . .

. . . .

. . . .

. . . .

. . . .

QUESTION 8 What are the relative merits of a pyro-electric vidicon system composed with a cooled thermal imager?

Answer . . . .

. . . .

. . . .

. . . .

. . . .

QUESTION 9 Does thermal imaging meet the requirement for a battlefield surveillance system?

Answer ....

....

....

....

ANSWERS ON PAGE 191

5.

Lasers

INTRODUCTION

Since laser action was confirmed in 1960 it has found use in a wide range of applications in the defence, industrial and medical fields. Laser radiation is distinguished from ordinary light in several respects: the most important is that it can be produced as an almost parallel beam which is also very intense and monochromatic. The beam can be directed with very high accuracy and intensity, which makes the laser a very important tool in application to surveillance and target acquisition in the military sphere, particularly in range finding, target illumination, designating and tracking roles. The aim of this chapter is to describe the principles of laser operation, the different types of laser which are being deployed in these operations, and the applications themselves.

PRINCIPLES OF LASER OPERATION

Basic Elements

The word LASER is an acronym for Light Amplification by Stimulated Emission of Radiation and is usually associated with emission in the near ultra-violet, visible, and infra-red regions of the electro magnetic wave spectrum. Optical stimulated emission is produced from an active material in which an inversion of the population of natural energy states has been contrived. The active material is bounded by reflecting surfaces between which the laser light is contained within a resonant cavity in similar fashion to the microwave generator. The active element can be either a gas, a doped crystal, a semi-conductor or a liquid and the population inversion may be obtained by pumping either by means of electrical, optical or electron beam methods. The laser output can be produced as either a continuous beam (CW mode) as a train of pulses (multiple pulse mode), or as a single giant pulse (Q-switching mode), of a few nanoseconds duration. These characteristics are discussed in more detail in the following sections.

Spontaneous Emission

In order to understand the differences between laser radiation and ordinary light it is necessary to examine their different mechanisms of formation.

Ordinary light, as from an incandescent lamp, is emitted by an excited atom through the release of an optical photon in the absence of an external stimulus. The process occurs at random and is called spontaneous emission. A photon is a packet or quantum of radiation which lasts for about 10^{-8} sec for an isolated atom.

Figure 5.1 illustrates the sequence of events occurring in spontaneous emission. An atom is raised to an excited state (a) by the absorption of a photon and, after staying in this state (b) for an arbitrary time, emits another photon (c) of energy $\Delta E = E_2 - E_1 = h\nu$ where ν is the photon frequency and h is Planck's constant. There is no amplification in this process.

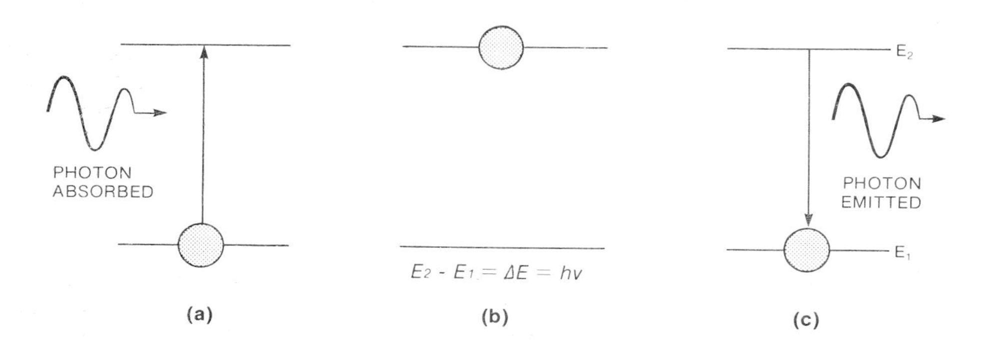

Fig. 5.1 Spontaneous emission

Stimulated Emission

The unique qualities of laser radiation are coherence, monochromaticity, directivity and high irradiance. They arise from the process of stimulated emission which is illustrated in Fig. 5.2.

It was shown by Einstein that if an atom or molecule is in a higher energy state the release of this stored energy can be controlled by subjecting the atom or molecule to an electro magnetic field of the same frequency. Comparison of Fig. 5.2 with Fig. 5.1 shows the essential difference between stimulated and spontaneous emission. For stimulated emission to occur the atom or molecule must already be in an excited state before it absorbs a second photon (Fig. 5.2 (c)) and before it has time to decay by spontaneous emission. As a result a second photon is emitted in the same direction and with the same frequency and phase as the stimulating photon.

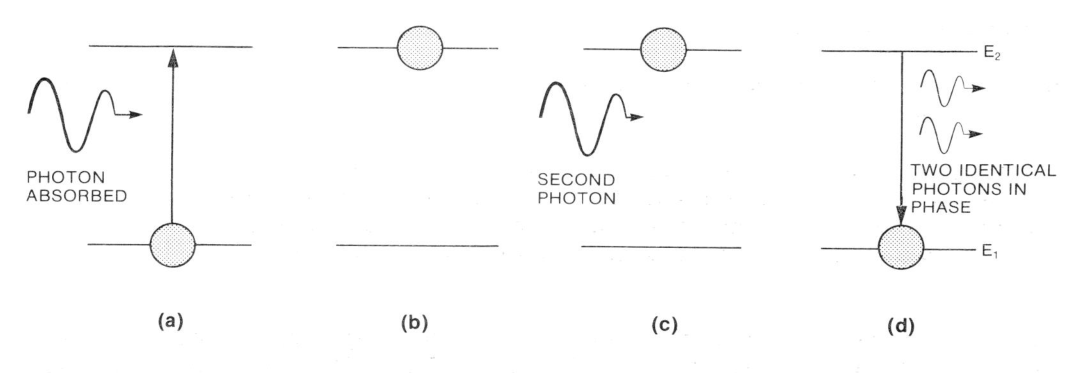

Fig. 5.2 Stimulated emission

Coherence

The stimulated photon is said to be coherent with the stimulating photon and they reinforce one another to produce an amplified, steady, and highly directional output wave. This is in contrast to the random, noisy, and isotropic radiation which arises in the spontaneous emission from an incoherent source. Practical lasers emit a narrow band of frequencies, thus coherence is not total and results in the output beam having a small divergence.

Cavity Resonator

For laser action to be sustained there must be a greater population of atoms in the higher energy state E_2 than in the lower energy state E_1. This is the opposite to the condition which occurs in the natural state of matter and is called population inversion. A cavity resonator is used to produce population inversion and also to amplify the wave. There is a close analogy with the microwave generator, the main difference being that there are many more competing modes in the optical cavity making a high degree of coherence more difficult to achieve. In its simplest form, Fig. 5.3, it consists of an active medium between two plane mirrors one of which is partially transparent to allow an external output in the form of a laser beam.

Pumping is achieved by external excitation which in the illustration is by electrical discharge into a gas medium. The pumping energy is used to raise atoms or molecules into the higher excited states and achieve a state of population inversion. Laser action is initiated by photons which are emitted spontaneously after pumping is initiated. These photons interact with the excited atoms as they pass through the cavity causing them to lose energy by stimulated emission. Each photon produces a wave which grows in amplitude on successive interactions with the gas atoms and this wave is reflected back and forth by the cavity mirrors stimulating more atoms producing further amplification. If the gain by stimulated emission exceeds the losses which occur for example by scattering and absorption in the active medium and mirror reflection losses, the process continues. There is of course a deliberate loss through the partially reflecting mirror and good

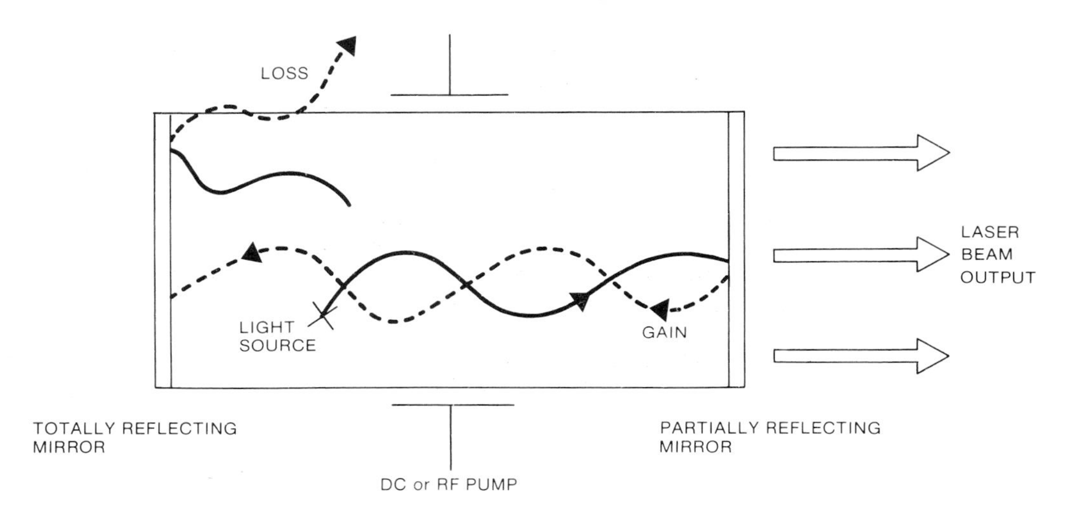

Fig. 5.3 Concept of cavity resonator

laser design is concerned with optimising the output coupling. If this is too high the loss rate may exceed the pumping rate and the oscillation will cease and if too low there will not be sufficient output power. Generally speaking a mirror transmission of 1-10% is usually employed. The efficiencies of most laser systems, measured as the ratio of the radiated coherent power to the electrical input power lie in the range 0.05 to 20%, the remainder being dissipated in non-collectable energy by photons moving non-paraxially in the cavity and by processes of absorption and spontaneous emission.

LASER BEAM PROPAGATION

The diameter D of the laser beam increases with range R according to the following equation:-

$$D^2 = d^2 \left(1 + \frac{\lambda R}{\pi d^2}\right)$$

where λ is the wavelength of the laser radiation and d is the diameter of the beam at the cavity. Table 1 shows how this varies for a neodymium laser.

TABLE 1 Laser Beam Diameter in cm as a function of Range

Laser output diameter (cm) \ Range (m)	10	10^2	10^3	10^4
0.2	0.6	6	60	600
2	2	2	6	60
10	10	10	10	12
20	20	20	20	20

The distance over which the laser beam is parallel is proportional to the square of the cavity diameter and inversely proportional to the wavelength. The divergence beyond this distance is proportional to the wavelength divided by the diameter of the laser cavity. As these same relationships apply generally for electromagnetic radiation it can be seen that a microwave generator operating at 1 cm wavelength would require a 500 cm diameter transmitter to produce the same range of beam parallelism as a neodymium laser with a cavity diameter of 5 cm. However the laser beam divergence beyond this range limit would be one hundred times smaller for the laser. The energy of the laser beam is thus concentrated in a smaller area and this high directionality gives the laser beam the potential for operation over much longer distances than the microwave transmitter. The tight collimation of the laser beam makes it difficult to detect and it has therefore potential deployment in covert operations though it could be countered by electro-optic means.

In some military applications it is important to achieve the smallest beam diameter at the target. In these cases the laser beam is passed through a telescopic beam expander. This acts to converge the output beam to a degree which is determined by the diameter of the final focusing optic, the limit being set by diffraction effects. It would of course be impossible to produce such small spots with a natural, incoherent light source without unacceptable sacrifice in intensity.

Atmospheric Effects

In a vacuum, as in space, the laser beam intensity is reduced only by its divergence and by the degree of beam jitter which is present. The beam divergence is proportional to the wavelength, thus the intensity varies as the inverse square of the wavelength, indicating the advantage in using short wavelengths in such applications. This relatively simple situation does not hold for terrestrial conditions because the laser beam interacts with the constituents of the atmosphere.

The main sources of intensity loss in passage through the atmosphere are:

1. Through absorption and scattering by the air molecules and aerosols which lie within the laser beam path. There are several types of aerosols such as water droplets which are encountered in haze, mist, fog, cloud and rain conditions. In addition there are solid particulates such as dust, fumes and smokes. Absorption and scattering do not materially contribute to beam broadening.

2. Through atmospheric turbulence, which produces temperature variations and variable refractive indices, causing intensity fluctuations and beam steering problems as well as beam broadening.

3. Through beam jitter created by inconsistencies in laser output and in servo-controlled tracking systems if they are used.

4. Through the beam divergence which is determined by the degree of coherence.

The laser energy which is absorbed by the air molecules and aerosols heats the local air and, if large enough, can produce density and refractive index gradients appropriate to a negative power lens. The effect is to defocus the laser beam and the term "thermal blooming" has been coined in analogy with a flower. Thermal blooming is an important source of intensity loss for high power laser transmission but can be reduced or eliminated by transverse winds and slewing.

Wavelength Dependence

The transmission of visible and infra-red radiation through the atmosphere is an irregular and complex function of the wavelength as is illustrated in Fig. 4.2. It can be seen that there are regions of high transmission, but also regions of strong absorption which is due principally to carbon dioxide molecules and water vapour.

It is of course the transmission windows which are of practical value. These occur in the visible; in the near infra-red region between 0.4 μm and 2 μm, but with some molecular absorption bands between 0.9 μm and 2 μm; in the middle infra-red region between 3 μm and 5 μm except for a large carbon dioxide absorption band at 4.3 μm; and the whole of the far infra-red region between 8 μm to 14 μm. This particular transmission spectrum relates to an actual coastal horizontal path and although the general characteristics hold for other geographical locations it can be expected that the presence of industrial smokes and fumes will have an important effect. Diurnal variations associated with local terrain conditions can also be expected, making prediction difficult.

Generally speaking it is aerosol scattering which dominates in the visible and near infra-red regions of the spectrum whereas molecular absorption is the important loss mechanism for the far infra-red region. There is an order of magnitude increase in penetration through aerosols for CO_2 radiation over visible radiation for visibility ranges in excess of about 5 Km. Under such conditions the droplet sizes are probably less than 1 μm. For fog conditions however where the droplet sizes increase up to about 50 μm the long wavelength advantage in aerosol penetration is lost. The total attenuation is better for the longer wavelength with a suggestion of an optimum wavelength at the top of the middle infra-red transmission band, but seasonal and diurnal variations in the atmosphere conditions play a critical role.

Table 2 gives an indication of the sizes of some of the important atmospheric constituents. It is to be expected from this data that increased range effectiveness of the middle and far infra-red bands would hold for dusts and smokes as well as haze but not for heavy fumes and fogs.

TABLE 2 Aerosol Sizes

Smokes	0.2-2 μm
Dust	1-10 μm
Fumes	Up to 100 μm
Haze	Up to 1 μm
Fog and Cloud	5 to 50 μm
Mist	50 to 100 μm
Drizzle	100 to 500 μm
Rain	500 to 5000 μm

Atmosphere turbulence causes beam broadening and wander to an amount which depends upon the range, the degree of turbulence and the wavelength of the laser radiation. Typically high turbulence could produce an order of magnitude increase in beam diameter at 5 μm. The wavelength dependence is very weak, but there is a critical wavelength below which turbulent beam spreading dominates over diffraction effects.

Because molecular absorption is stronger at the longer wavelength it is expected that thermal blooming will be the intensity limiting feature for high power propagation whereas in the visible and near infra-red region turbulence induced beam spreading and aerosol scattering are the limiting propagation effects. The middle infra-red region appears to offer the best compromise though the actual physical and atmospheric conditions will again determine the optimum wavelength.

Return Signal

The average power P_r which is collected by a rangefinder receiver of area A_r from a target at range R which diffusely reflects the whole of the laser beam energy with a reflection coefficient ρ is given by:-

$$P_r = \frac{\rho Po}{\pi} \frac{A_r}{R^2} \eta_t \eta_r \exp(-2\mu R)$$

where Po is the average radiated power and η_t and η_r are the quantum efficiency coefficients of the transmitter and reflector respectively.

The return signal strength must be greater than a minimum value which is determined by the noise characteristics of the receiver detector. The signal to noise ratio S/N is related to these characteristics and to the return signal strength by the following equation:

$$S/N = \frac{P_r}{((NEP)^2 \times 4/\tau)^{\frac{1}{2}}}$$

where NEP is the Noise Equivalent Power of the detector and τ is the pulse width.

By way of example to illustrate these features:

If Po = 1 megawatt, τ = 40 nanosec, the receiver aperture diameter is 5 cm and the wavelength is 1.06 μm (neodymium) the return signal from a target of reflectivity 0.01 at a range of 1 Km is approximately $0.5\ 10^{-5}$ watts making due allowance for atmosphere losses over the outward and return paths. For an assumed NEP of 10^{-10} watts $(Hz)^{-1}$ the signal to noise ratio S/N would be about 5 to 1 and the target could easily be ranged. Here $\eta_t = \eta_r = 1$.

If the target does not intercept the whole of the laser beam the expression for P_r above should be multiplied by $\frac{4A_t}{\pi R^2 \theta^2}$ where A_t is the area of target and θ is the laser output divergence.

In surveying and co-operative applications a corner cubic reflector is suually fitted to the target. This increases the return signal some 10^4 times stronger than from a diffusely reflected target like an enemy tank and greatly increases the effective range.

TYPES OF LASERS

Pumping Cycle of a Typical Laser

The pumping action in a practical laser is more complicated than the simple system described earlier in the chapter, and more than two atomic levels are usually involved. The pumping oscillation cycle of a typical laser system which operates at four levels is shown in Fig. 5.4.

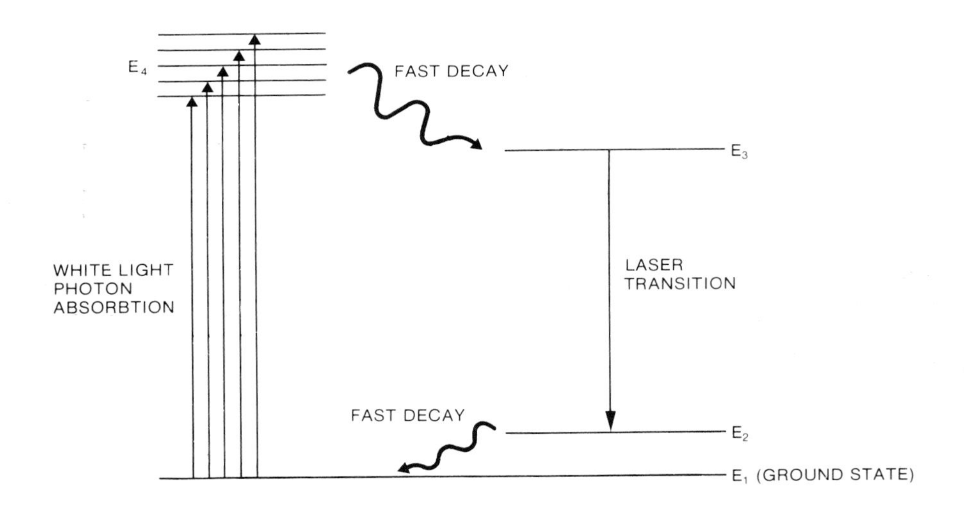

Fig. 5.4 Pumping cycle for a 4-level laser

In such a system the highest energy state E_4 is a broad absorption band of closely spaced levels. These make an efficient reservoir for light of different frequencies (white light) to raise electrons from the ground state and the principle is used in the ruby and neodymium lasers. The excited atoms will quickly decay to level E_3 followed by a laser transition to E_2 and subsequent fast decay to the ground state E_1. A similar process occurs in a gas laser; the main difference is that the absorption is taken up by one type of atom which then transfers energy by collision with atoms of a second kind and these atoms experience a laser transition. In either case a condition of population inversion is strongly established, particularly if the lower state of laser transition E_2 is not also the ground state E_1.

Gas Lasers

Gas lasers are simple in construction and offer the greatest degree of versatility. They are capable of achieving a high degree of coherence and directionality, and cover the widest range of output power.

The helium neon laser was the first gas laser to be demonstrated. Optical pumping is achieved by direct-current or radio-frequency excitation of the helium atoms and energy is transferred to the neon atoms by atomic collision. The neon atoms radiate by laser action principally at 0.6328 μm but other lines are emitted as well. Power outputs range from 1 mW to 20 mW and uses are mainly associated with metrology and surveying range-finding. A schematic diagram of a helium neon laser is shown in Fig. 5.5.

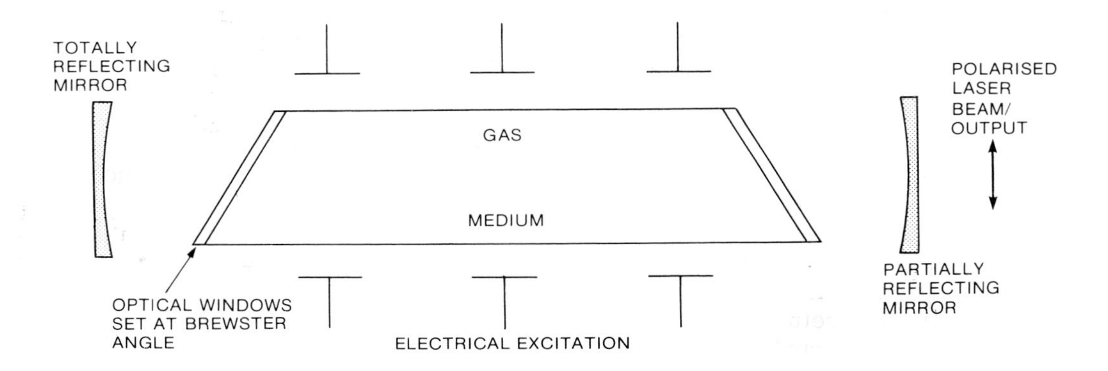

Fig. 5.5 Helium Neon Laser

The cavity is usually decoupled from the mirrors to protect them from damage and the tube is sealed by a pair of anti-parallel optical flats set at the Brewster angle, thus the laser output is polarised. Beam quality is usually very good but efficiency is less than 1%.

The argon laser operates on the basis of transitions between excited states of singly ionised argon atoms which emit on several lines between 0.35 μm and 0.53 μm. As it takes much more energy to ionise each atom more input is needed for an argon laser compared with a helium neon laser of the same output and water cooling may be required, adding bulk.

The principal gas laser of current interest to surveillance and target acquisition is the carbon dioxide laser. This works by molecular transitions in the carbon dioxide molecule which are closer together than atomic states and therefore a lower laser photon energy results ie a longer photon wavelength is emitted. The carbon dioxide molecule emits in the range 9 to 11 μm with the strongest line at

10.6 μm. This is in the middle of the far infra-red atmospheric transmission window. Nitrogen and helium gas are also present. The nitrogen has the same role as does helium in a helium neon laser. The nitrogen molecule is excited by an external "pumping" source and exchanges energy with carbon dioxide molecules by collision. The helium assists population inversion by breaking down some of the non-productive levels in the carbon dioxide molecule. Efficiencies of 20% have been achieved in low pressure continuous wave CO_2 devices and output powers well in excess of a kilowatt have been obtained.

The difficult problems of achieving reliable pulsed operation have been overcome by the development of the Transversely Excited Atmospheric (TEA) Laser. This laser produces stable and uniform conditions of discharge in the laser medium by a carefully designed configuration of electrodes in the laser cavity.

Peak output powers in the region of 200 KW over a pulse length of 60 nsec have been achieved with a cavity length of 25 cm and with an efficiency of about 1%. The TEA laser forms the basis of the Ferranti and Marconi laser range finders which are briefly described later. The output is polarised in a similar manner to the helium neon laser just mentioned.

Development in the use of the CO_2 laser for ranging applications is likely to be in the use of high repetition rates for combined ranging and target designating purposes. Very compact high pressure waveguide type lasers of the continuous output type are also being developed for communications.

Another growth area is the use of continuous wave CO_2 lasers with heterodyne detection. The principle of operation is to frequency modulate the continuous output, and mix the return signal, which may be frequency shifted, with a component of the modulated transmitted signal to produce a beat frequency which is detected directly. Target range and radial velocity information is obtained from the output signal. Such systems are the infra-red analogue of microwave radar and the concepts and different modes of operation of the conventional technique are being explored in the laser wavelengths. The technique avoids some of the problems of sensitivity which are encountered with direct detection in the infra-red, and higher signal to noise levels can be achieved. The technique can be used in the co-operative roles of ranging and communication and is probably less detectable than the direct pulsed method because the emitted energy is lower. The main disadvantage is that the signal processing is more complex and this leads to bulkier equipment and less portability.

Other gas laser developments which may have relevant applications are the chemical and excimer laser types. The chemical laser works on the basis of combustion of fluorine with hydrogen or deuterium gas to produce emission at 2.5 μm and 3.8 μm respectively, the latter coming within the middle infra-red atmospheric transmission window. The excimer laser uses a mixture of rare gas and halogens to lase in the visible and ultra-violet. Examples are krypton fluoride emitting at 0.25 μm and argon fluoride at 0.19 μm. Efficiencies in excess of 10% appear to be possible.

Solid State Lasers

Solid state lasers offer the advantage over gas lasers of having a much higher volume density of ions and this makes for the smallest system and portability. Most of the laser range finders in use today use an optically pumped solid state laser. Because they are so small and have low conversion efficiency heat dissipation is a problem and they are not used for high average power applications.

The principle of operation of the ruby or neodymium laser with an elliptical cavity is shown in Fig. 5.6. Since pumping by atom collisions is not possible in solids the laser material must be optically transparent to the radiation from the pumping source which is a white light Xenon flash tube. The flash tube is placed at one focus and the ruby or neodymium rod at the other. The ruby crystal consists of aluminium oxide to which has been added a small proportion of chromium and it is the chromium in the form of ions which provides the laser transition levels.

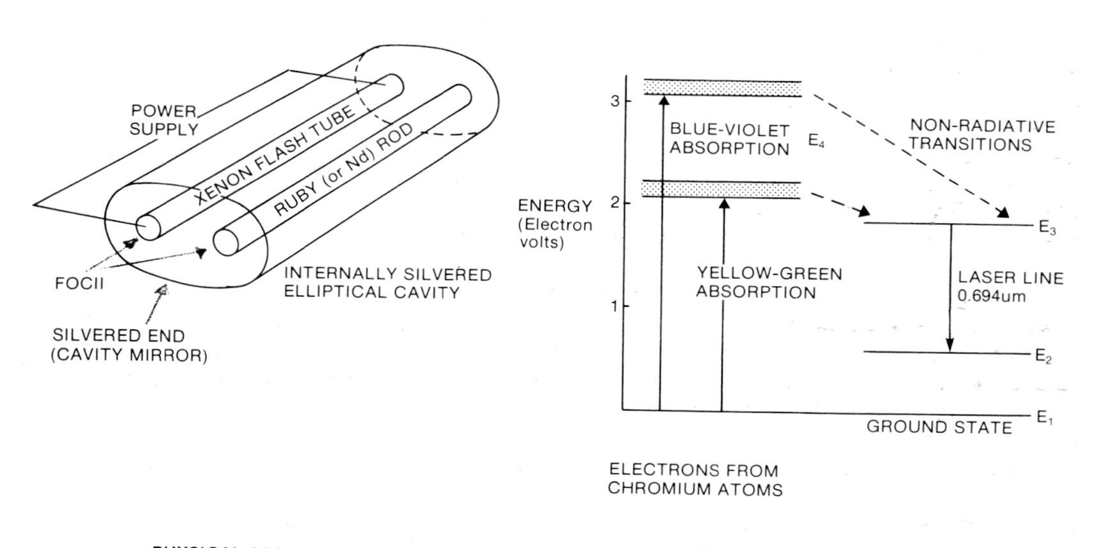

Fig. 5.6 Principle of operation of the ruby and neodymium laser

The ruby crystal suffers from low efficiency (0.05%), causing heating at very low pulse repetitive frequency which quenches the pumping action. It is necessary to cool the crystal to liquid nitrogen temperature in order to maintain a reasonable pumping speed. Although the ruby laser is the original equipment for the Chieftain range finder, later service range finders use neodymium with either yttrium aluminium garnate (Nd:YAG) or glass (Nd:glass) as the host material. The laser output for the neodymium laser is 1.06 μm and its efficiency is about 2%.

The cavity oscillator is usually formed by silvering the ends of the laser rod or by using separate cavity mirrors. The output from such a simple resonator is a mixture of transitions between various transverse and axial modes which can escape from the system and the resulting temporal shape is a series of spikes from these competing modes of oscillation. In order to make this suitable for military applications the output pulse must be controlled, and in the case of current ranging systems this has been achieved by a technique called Q-switching.

Q-switching involves changing the quality of the cavity. Initially the cavity mirrors are removed and pumping allows the population inversion to build up to a very high value without the feedback which would have built up the competing oscillations. After the population inversion has achieved the highest possible value the mirrors are switched back into the system suddenly increasing the Q-factor and feedback, allowing the whole of the stored energy to be emitted in one short, very intense pulse, essentially of a single mode oscillation.

Q-switching can be done in several ways. In one technique a mirror or porro-prism is rotated at high speed eg 30000 rpm and only when this is aligned with the stationary mirror will the energy be released as laser output (Fig. 5.7). This type of device produces pulses of width 50 nsec but the switching speed is relatively slow.

A faster method employs an electro-optically generated switch, which is also shown in Fig. 5.7.

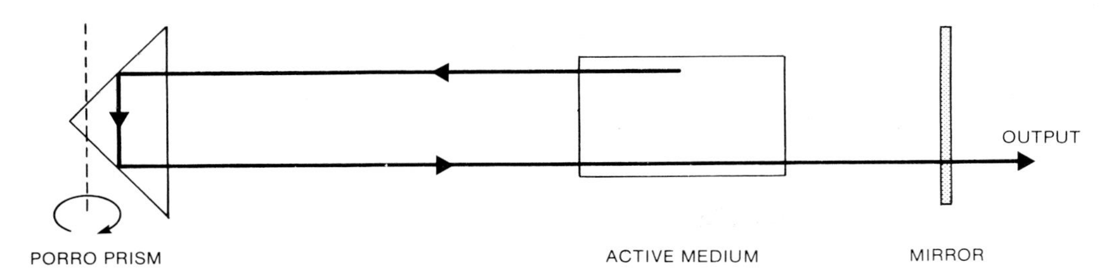

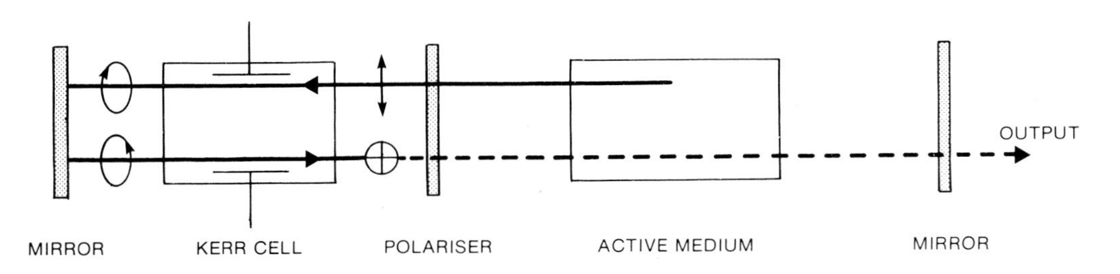

Fig. 5.7 Two types of Q-switch

A Kerr-cell contains nitrobenzene placed between a pair of electrodes. When an electric field is applied an optic axis is induced parallel to the direction of the electric field. Only along the optic axis is the refractive index independent of the direction of polarisation of light. Plane polarised light incidence on the cell can in general be resolved into a component which vibrates parallel to the optic axis and one component which vibrates perpendicularly to it. The velocities of the two components within the cell will be different and on emerging from the crystal will be out of phase and generally output will be elliptically polarised. If the angle of incidence of the incident plane polarisation is 45° to the optic axis and the phase difference is a multiple of $\pi/2$ the output will be circularly polarised. In Fig. 5.7 the cavity output light is made plane polarised by the polariser and converted to circularly polarised light by the Kerr-cell. The cavity mirror reflects this polarised light and in doing so reverses the direction of polarisation. Thus the light which re-emerges from the Kerr-cell is also plane polarised but at right angles to that transmitted by the polariser, thus it cannot be transmitted again by the polariser until the electric field is switched off. With the electric field on, the population inversion is allowed to increase and the mirrors are "switched in" when the field is reduced to zero. The change in voltage must be synchronised with the pumping. With this technique output pulses of 10 nsec can be produced.

For higher gains a saturable dye is introduced into the laser medium. Normally the dye strongly absorbs at the laser wavelength so that no amplification is possible. A point is reached in the build up of the population inversion when the gain due to stimulated emission exceeds the losses and laser action begins to take place. The dye becomes saturated by the optical flux and is very rapidly switched into a state of transparency. The stored energy in the very large population inversion is immediately emitted as an optical pulse in a few nanoseconds. The characteristics of the dye, eg its absorption cross-section and time constants, have to be carefully matched with the host material. This photo chemical method is very suitable for lightweighting applications.

Another method of controlling the output pulse is to use the technique of cavity dumping. Here the population inversion is allowed to build up to a peak value as in Q-switching at which time an electro-optic switch is opened to allow the stored energy to be dumped into the cavity as optical radiation. Little or no radiation is allowed out of the cavity until the whole of the energy is converted, and at this time the switch opens a second channel so that all the optical energy is emitted in pulse times of the order of 1 nsec. With these techniques ranging accuracies of about 10 cm can be achieved.

An even faster technique is called mode locking. The transmission of the electro-optic switch is modulated and synchronised to correspond with the photon round trip time in the cavity and enables pulses in the picosecond region to be produced.

Other types of solid state laser which are being considered are holmium in yttrium lithium fluoride as the host material (HO YLF) which emits at 2.06 µm and erbium in glass which emits at 1.54 µm. Both of these laser wavelengths are beyond the maximum corneal transmission and are regarded as relatively eye-safe. The problem is that neither material is as efficient as Nd:YAG and much more input energy is required to drive them.

Semi-Conductor Injection Lasers

These systems work on the principle of a p-n junction biased in the forward direction so that the positive holes are injected from the p-region towards the n-region and recombine with electrons in the junction to give up their energy as heat or light. Figure 5.8(a) illustrates this principle and Fig. 5.8(b) the physical arrangement.

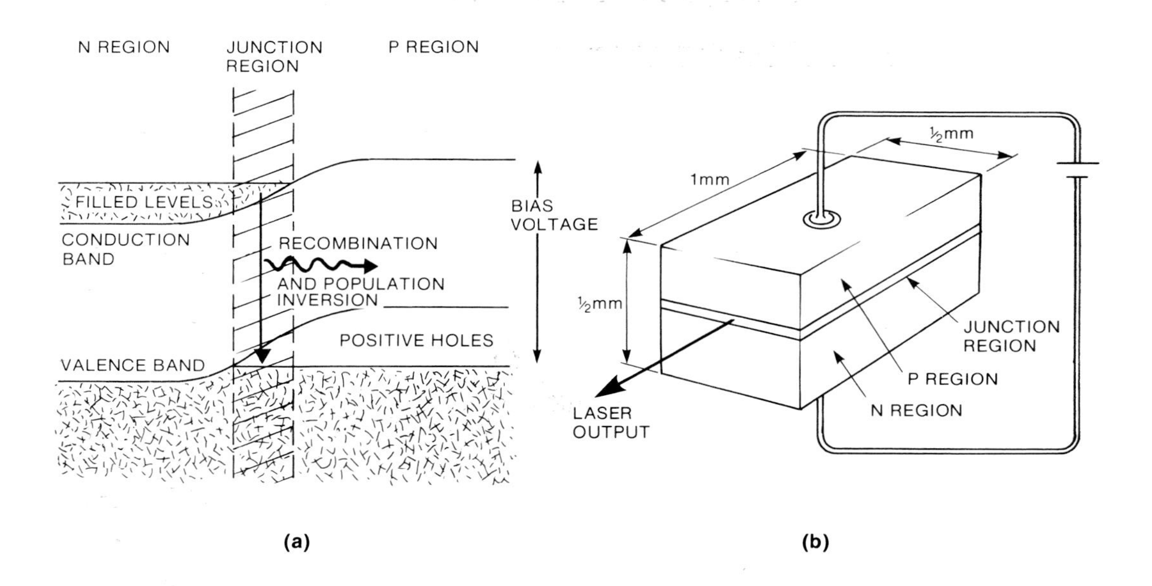

Fig. 5.8 Semi-conductor laser

At high current densities, near 5000 amps cm^{-2}, the p-n junction in GaAs behaves as if it had a negative absorption coefficient for radiation so that when radiation passes through the junction region it is amplified; and if it is made in the form of an optical resonator the system becomes a laser.

Outputs of several milli watts are achieved at efficiencies up to 50% in a wavelength band between 0.84 μm and 0.90 μm. Modulation of the output provides the facility for ranging and communications.

The particular attractions of GaAs lasers are that they are highly efficient photodiodes of compact size and can be directly modulated by their drive current supply. They are however of low power and large beam divergence and thus have limited range finding application, probably to 100 m.

LASER SAFETY

Protection Standards

The cornea of the eye is transparent to optical radiation in the range 0.45 μm to 1.4 μm. Laser radiation in this region is focused on the retina which is of course very sensitive and therefore easily damaged. Table 3 is an extract from the UK Defence Standard 05-40 Issue 2 of 1977 which defines Protection Standards for different laser wavelengths of pulsed and continuous wave types.

TABLE 3 Protection Standards

(Extract from UK Defence Standard 05-40 Issue 2 1977* which should be referred to for precise safety information. This Table should not be regarded as official safety advice but merely to show relative effects).

Type of Laser	Output Profile	Wavelength μm	Exposure	Protection Standard at the cornea
Ruby	single pulse	0.694	1ns-18ns	5 10^{-7} J cm^{-2}
Neodymium	single pulse	1.06	1ns-100 s	5 10^{-6} J cm^{-2}
Gallium Arsenide	single pulse	0.85	1ns-18 s	10^{-6} J cm^{-2}
Holmium	single pulse	2.06	1ns-1 s	1 10^{-2} J cm^{-2}
Carbon Dioxide	single pulse	10.6	1ns-1 s	1 10^{-2} J cm^{-2}
Helium Neon	CW	0.63	0.25s	2.5 mW cm^{-2}
Argon	CW	0.5	0.25s	2.5 mW cm^{-2}
Neodymium	CW	1.06	100s	0.5 mW cm^{-2}
Carbon Dioxide	CW	10.6	10s-100s	0.1 W cm^{-2}

J = Joules W = watts

The Protection Standard is defined as the intensity of laser radiation at the cornea, measured in J cm^{-2} per pulse for pulsed lasers and W cm^{-2} for continuous wave lasers, below which exposed persons will not sustain eye injury.

From this Table can be seen the extremely low Protection Standard for the pulsed ruby laser and the improvement by a factor of 10 for the neodymium pulsed laser. The Protection Standard for pulsed holmium and carbon dioxide lasers is 1 10^{-2} J cm^{-2}, a further improvement of 2000 over neodymium. This is because the laser radiation is all absorbed in the cornea which has a much higher damage threshold than the retina.

The same advantages in the use of longer wavelengths is apparent for continuous wave lasers. The Protection Standard for carbon dioxide is 0.1 W cm^{-2} which is only a factor two higher than ambient blackbody conditions.

*To be revised.

Nominal Ocular Hazard Distance

The Defence Standard identifies a Nominal Ocular Hazard Distance (NOHD) for range operations and is defined as the distance at which the laser intensity has fallen to the level of the appropriate Protection Standard. It is given by the following expression:

$$\text{NOHD} = \frac{\sqrt{\dfrac{1.27\,Q\,(\tau M^2)}{P_s}} - a}{\phi}$$

where

Q = laser energy (J) or power (W)

P_s = protection standard (J cm^{-2} or W cm^{-2})

ϕ = beam divergence (radians)

a = emergent beam diameter at $\frac{1}{e}$ of peak intensity

τ = transmission of optical instrument if used

M = magnifying power of optical instrument if used

NOHD's for some of the laser range finders of interest are given in Table 4.

TABLE 4 Nominal Ocular Hazard Distances

Laser	NOHD
Chieftain Ruby LE2	10 Km
SIMRAD LP7	3 Km
LRMTS	20 Km
CO_2 Laser (2Watts)	zero range

A further safety factor which should be taken into account is that it is necessary to allow for 'hot spots', or scintillation, in the beam. A multiplying factor of 10 is made to the intensity, increasing the NOHD by $\sqrt{10}$ to define an Extended Nominal Ocular Hazard Distance (ENOHD).

Further correction factors are prescribed for repetively pulsed lasers having pulse durations of less than 10 μsec; eg in the case of the Laser Ranger and Marked Target Seeker (LRMTS) the protection standard is reduced by a factor of 3. Other factors take into account the differing sensitivities of the retina for wavelengths between 0.7 μm and 1 μm.

A laser hazard classification system groups lasers into four classes according to their hazard effect. Class I lasers have no restrictions because their NOHD is effectively zero, Class II and III lasers require protective goggles whilst Class IV lasers emit radiation which is dangerous to eyes and skin even after reflection from a diffuse surface and must be totally contained. Proprietary goggles are available marked with the attenuation they provide against the laser radiation wavelength from which they are designed to protect.

APPLICATIONS TO SURVEILLANCE AND TARGET ACQUISITION

Uses

This section describes the principles of some of the in-service equipment being produced for ranging, designating, illuminating and tracking purposes and also a laser based direct fire simulator which is used for gunnery training.

Range Finding Principles

The laser range finder has significantly improved first strike operations at long distances. In operation, the laser range finder emits an intense, highly collimated beam of short pulses of light which are reflected by the target back to the receiver. The time taken for a single pulse to travel out to the target and return is measured and converted into a direct reading of range.

For ranging applications it is important to achieve high peak power output with narrow pulse width, because this combination gives long detection range and good range resolution.

Most of the laser range finders in operation at the present time use Q-switched solid state lasers. Such lasers give a range resolution of about $\pm$5m up to ranges of 10 Km within a beam divergence of 0.5 mrad.

Figure 5.9 shows the layout of a typical laser range finder. The laser output is coupled to a small transmitting telescope which decreases the cavity output divergence and hence increases the range. Part of the output is fed to a photodiode which provides the start pulse for a range counter. The laser beam is reflected by the target and some of the energy is collected by a receiver lens which is focused onto the detector. The detector may be protected by a narrow band interference filter which reduces the background noise. The detector output is amplified, filtered, and if above an acceptable threshold value, is used to stop the range counter. The number of pulses in the range display is a measure of the range and this is converted into a direct reading.

The operation of the range finder is made as simple as possible. In a typical case the operator sights through an optical telescope and centres the target in the field of view. He then depresses a fire button which activates the transmitter pulse and the range is immediately displayed in the bright field of view. Other information also appears in the field of view such as the state of charge of the battery. In some cases circuit logic is able to distinguish between several potential targets by range gating which is also displayed to the operator.

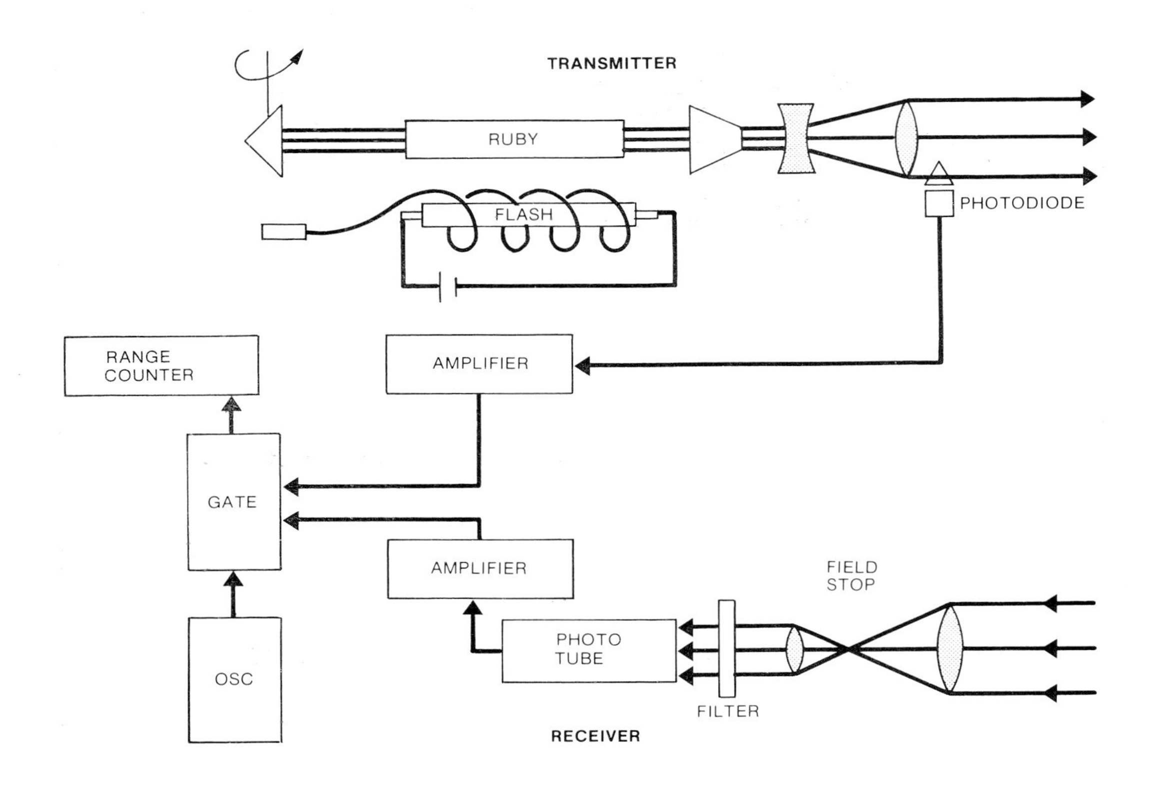

Fig. 5.9 Principle of pulsed laser range finder

Equipments are in service for range finding prior to engagement with indirect fire weapons or as integral part of direct fire weapons such as tank guns.

Laser range finders of the pulsed type can be man portable weighing in the region of 3 kg but much heavier ones weighing about 30 kg, are fitted to tanks or helicopters.

Field Range Finders

The characteristics of two field instruments are summarised in Table 5.

TABLE 5 Two Field Type Laser Range Finders

Parameter	Units	Hand Held Range Finder SIMRAD LP7	Tank Laser Range Finder LF-2
TRANSMITTER			
Type (All Q-switched)	-	Nd: YAG	Ruby
Wavelength	m	1.064	0.694
Output power	MW	0.5	1
Pulse width	ns	8	40
Output energy	mJ	4	40
PRF	Hz	Single shot or continuous	Single shot
Beam diameter	cm	3	4.7
Beam divergence	m rad	2	0.5
RECEIVER			
Field of view	m rad	1.3	3
Aperture	cm	4.5	4.7
Detector	-	Silicon avalanche photodiode	Photo mlutiplier
Min range	m	150	500
Max range	m	4000	10,000
Accuracy	m	± 10	± 10
SIGHT			
Field of view	m rad	120 (7°)	153 (8.5^{o})
Magnfiication	-	x 7	x 10
POWER SUPPLY			
Type	-	12v, 0.4Ph Ni Cd battery	28v vehicle supply
Number of shots	-	600	10 shots per minute
WEIGHT			
Range finder and battery	kg	1.7 (3.7 lb)	36 kg total weight

The Barr and Stroud LF-2 range finder was developed for use on the Chieftain tank. As originally designed the system uses a ruby laser with a spinning porro-prism Q-switch to produce an output pulse of 40 nsec, but a passive dye Q-switched Nd:YAG version is now available with a 10 nsec pulse width. Further particulars of the ruby system are given in the table.

The LF-2 range finder is aligned to the axis of the tank main gun and the sight is laid onto the target by the gunner using a ballistic graticule. The range of the target is presented to the gunner on the left hand eyepiece and he uses this information to elevate the sight and gun axis until an aiming mark, selected according to the range and type of ammunition in use, is aligned onto the target. The tank gun is then set to produce the correct ballistic trajectory to the target. In some versions the sight is coupled on line to a computer for direct fire control applications. In this mode the computer controls a cathode ray which is used as an aiming mark and is presented into the gunner's sighting system, or it can be used in conjunction with a thermal imager to provide the gunner with a capability of ranging at night or in poor visibility.

The carbon dioxide laser is now receiving attention for range finding operations. It has several advantages over the ruby and neodymium lasers. Firstly, it has increased atmosphere penetration and is less sensitive to haze, mist and battlefield smokes. Secondly, the CO_2 range finder is optically compatible with thermal imaging devices which are now being developed for use in the far infra-red wave band and co-operative action, with common optics and detector, is highly probable.

A third advantage is that as glass is opaque to 10.6 μm the operators of conventional direct viewing systems, such as binoculars, are protected by the front optic. A further important advantage of the CO_2 laser is that the eye is much less sensitive to radiation of wavelength 10.6 μm than it is to visible or near infra-red light. This is because the retina is much more sensitive to damage than the cornea which does not transmit wavelengths greater than 1.4 μm.

A CO_2 laser range finder operating single shot with a 50 nsec pulse and a 5 cm diameter germanium lens aperture will not exceed the Protection Standard for corneal damage even at zero range for peak power outputs less than 5MW which is considered ample for ranging operations. A CO_2 continuous wave laser with a 5 cm diameter optic can operate up to 2 Watts output before exceeding corneal damage threshold and this is also an ample power for ranging applications. These lasers have an NOHD of zero and are truly eye-safe.

Carbon dioxide laser range finders are being developed by Ferranti and by Marconi Avionics. Each is based upon the Transversely Excited Atmospheric (TEA) laser. Ferranti have developed and tested two pulsed types; one weighing 16 kgm producing 350 kW peak power output over a 15 cm aperture with 60 nsecs pulse length and maximum range 6 km, which uses a Lead Tin Telluside (LTT) receiver detector operating at 80°K by a Joule Thompson compressed air cooling system. Another weighs 7 kgm and operates at maximum range 20 km with 600 kW peak power over an 8 cm aperture with 60 nsecs and using a Cadmium Mercury Telluride (CMT) detector. Marconi Avionics have developed and tested a 220 kW peak power device with 60 nsec pulse operating up to 9 km using an LTT detector. Each type is accurate to ±5 km. (See Fig. 5.10).

Target Designation

The use of laser equipments is improving the effectiveness of close air support and of artillery. The principle of target designation is that the target is illuminated by a laser beam and a detector in the aircraft nose or artillery shell (so-called "smart bomb") homes in on the reflected light from the target. The laser's narrow beam assures very accurate, selective, marking accuracy at ranges up to

FERRANTI TYPE 307

MARCONI MARK III

Fig. 5.10 CO_2 laser range finders

about 10 km and reduces the chances of detection by the enemy. The designator can be operated by a soldier on the ground or it can be carried in a co-operative aircraft. The laser has great advantages over conventional radar in being able to provide very accurate ranging rate data at the shallow angles which are encountered in low level air attack of ground targets.

The complete system envisaged for British forces contains a Laser Target Marker and Ranger (LTMR) operated by a Forward Air Controller (FAC); and a compatible set of equipment in the aircraft, known as a Laser Ranger and Marked Target Seeker (LRMTS). The concept is illustrated in Figs. 5.11 and 5.12.

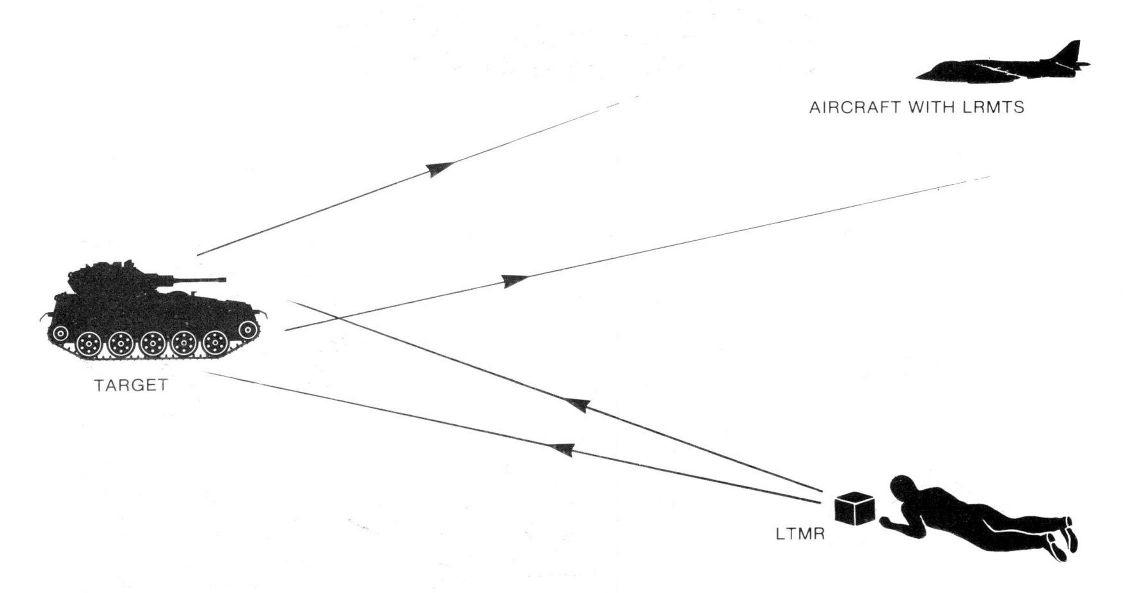

Fig. 5.11 Target designator concept

The principle of operation is that as the aircraft approaches the target the LTMR is initiated by the FAC and emits a stream of laser pulses which strike the target. Some of the reflected energy is detected by a quadrant detector in the LRMTS which activates gimbal servo motors to bring the sighting optic axis in line with the target and gives the pilot directional information enabling him to acquire and track the target automatically. The pilot flies the aircraft so that an aiming mark presented in his head-up display aligns with the target. Once the system has locked onto the target the LRMTS range finder takes over from the LMTR and gives the pilot a continuous update of the range to target. The pilot's task is then to use the ranging information and target datum in his head-up display to indicate the precise time for automatic weapon release.

If no LTMR is available the airborne equipment can be used in the ranging mode but the pilot's task is made more difficult. The LTMR can be used for a range of equipments which are being developed for homing onto laser illuminated targets.

a. LASER RANGER AND MARKED TARGET SEEKER IN THE HARRIER

b. LASER TARGET MARKER AND RANGER

Fig. 5.12 LTMR and LRMTS

A Ferranti LRMTS is in service with Jaguar, Harrier and Tornado. This is based upon an electro-optically Q-switched neodymium YAG laser which is capable of firing at pulse repetition frequencies in the range 10 to 20 Hz, the higher frequencies being particularly useful for operation with laser guided weapons such as PAVEWAY which is now entering service with the Royal Air Force.

The laser cavity is folded for compactness and high input stability, and the output is injected into a combined sighting and transmitter telescope. A silicon avalanche photodiode is used for ranging and a quadrant detector for posture control.

The Martin Marietta Corporation US have developed a laser guided artillery shell called Copperhead which is designed to destroy an armoured vehicle with one or two shots. Copperhead has a laser detector in its nose and has to survive very high linear and rotational acceleration, and also high temperature, when the gun is fired.

Target Illumination

The Army's requirement to enhance its night vision capability has led to the use of lasers to improve the performance of image intensifiers. In its simplest form such a device consists of a small torch which is used to illuminate the target when the ambient light conditions are unfavourable for the image intensifier acting alone. One such device uses a gallium arsenide laser in the continuous wave mode, having an output power of 100 mW giving a spot diameter which can be varied between 1.5 m and 8 m at 100 m range.

More sophisticated systems have been developed which use gated viewing of the target. This technique considerably reduces the effect of atmospheric backscatter which is the main cause of contrast degradation in conventional illumination of the target by incoherent near infra-red radiation. With a pulsed laser the receiver need be switched on only for a short time after each pulse is emitted and the effect of backscatter is greatly reduced. Viewing can either be direct by image intensifier or indirect by incorporating a television camera. The latter method is known as low light level television (LLTV).

Tracking

Lasers are most suitable for tracking targets particularly for low level and night-time operations. The laser range finder is often part of the tracking system. There are a wide range of tracking scenarios each one placing a different emphasis on the system. For example the requirements for tracking a satellite are not the same as those for tracking a missile or aircraft because the angular rates are vastly different; in this latter group there are different requirements for hostile and co-operative targets. In almost all cases however probably there will be a need for a conventional radar or optical viewer, of wide field-of-view to acquire the target, or targets, before handing over to the more accurate, but narrower field-of-view laser tracker. The analogy in astronomy is the use of a wide angle telescope to locate a stellar object, then changing to a high power narrow beam telescope to study its details.

Laser tracking systems operate on the same principles as do conventional radar but have important advantages over the microwave band. Because the laser wavelength is much shorter the size of the system is correspondingly smaller, and as the laser aperture is large compared with the wavelength a laser beam of near diffraction-limited output will have high directivity. The combination of good directionality with short wavelength make the laser system less sensitive to interference than in the microwave region where multi-path effects can be significant. Laser side lobe transmission is much less than for microwaves, making detection more difficult.

The disadvantages are that lasers are very sensitive to weathering effects particularly those which operate in the visible and near infra-red bands, and the problem of eye damage is more acute in this spectral region. Pulsed lasers are subject to photon noise effects which are not present at microwave frequencies whilst continuous working lasers require modulation and detection which add complications to the system.

Most types of lasers have been deployed in the various tracking systems. Solid state lasers such as ruby and Nd:YAG, and the semi-conductor GaAs laser have all been used in Q-switched pulsed mode; whilst CO_2, He-Ne, and Ar have been used in continuous working mode with external modulators.

The laser beam is usually expanded with a telescope in order to reduce the beam divergence and there are several ways of combining the laser with the telescope to direct the beam towards the target. An arrangement in which the beam is first expanded and then directed is called a coelostat, and, if the beam is directed before expansion the configuration is a Loudè telescope.

Some examples of laser tracking systems are outlined below. The reader is referred to more specialised articles for detailed descriptions.

A mobile van-mounted system has been developed by Sylvania and used for tracking co-operative aircraft. The system incorporates a Q-switched pulse neodymium YAG laser with output pulse of 50 mJ and width 15 nsec, which is capable of operating at a maximum repetition frequency of 100 Hz. The optical arrangement is a coelostat with the final transmitting mirror being adjustable in azimuth and elevation to give continuous pointing coverage of the target. Acquisition is via a joystick controlled vidicon television camera with near parallel bore sight and once the target is located in the television display the system goes into a fully automatic tracking mode. A retro-reflecting array is attached to the aircraft in order to enhance the optical return signal which is passed back along the same boresight as the transmitter through the transmitting mirror and to a beam splitter, thence to a ranging receiver and a tracking receiver. The latter incorporates a silicon diode quadrant detector which produces electronic signals for servo mechanisms to actuate the mirror for maintaining target station. The ranging receiver signal is used as a stop pulse in a range counter, measuring range in the same manner as was described earlier for the range finder. The system is capable of tracking to an angular accuracy of 100 μrad in azimuth and in elevation and can range to an accuracy between 0.2 m and 1.5 m depending upon the strength of signal. Maximum range is about 30 km and aircraft velocities up to 180 $msec^{-1}$ and angular rates up to 2 rad sec^{-1} are possible.

An example of a continuous wave laser tracking system is to be found at White Sands Missile Range in New Mexico. This has been used for tracking missiles from launch with a range accuracy of about 0.1 m and a tracking accuracy of 100 μrad. The system uses an argon-ion laser of 5W output which is modulated and transmitted through a fully adjustable mirror with a pointing accuracy of 50 μrad. The beam is sufficiently narrow to ensure sufficient return signal without the aid of retro-reflectors and this is applicable for engagement of hostile targets. The system is fully automatic and, unusually, acquisition is obtained by laser beam scattering over a field of view of ± 1 degree. As the aim point is directed at the missiles nose unwanted signal from the hot exhaust is much reduced.

A third system of a completely different type works on the Doppler principle which is illustrated in Fig. 5.13.

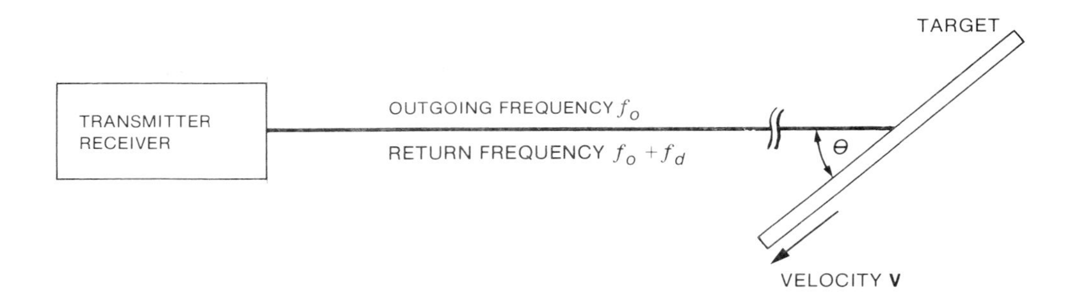

Fig. 5.13 Doppler principle

The frequency of backscattered radiation from a target moving with velocity V at an angle θ towards the transmitter is increased by an amount f_d where

$$f_d = \frac{2V\, f_o \cos\theta}{C}$$

and f_o is the frequency of the transmitter and C is the velocity of the laser radiation. A decrease in frequency will result for a target which is moving away from the transmitter.

This principle has been used to obtain coincident measurements of radial velocity as well as range of the tracked target, and is illustrated in Fig. 5.14.

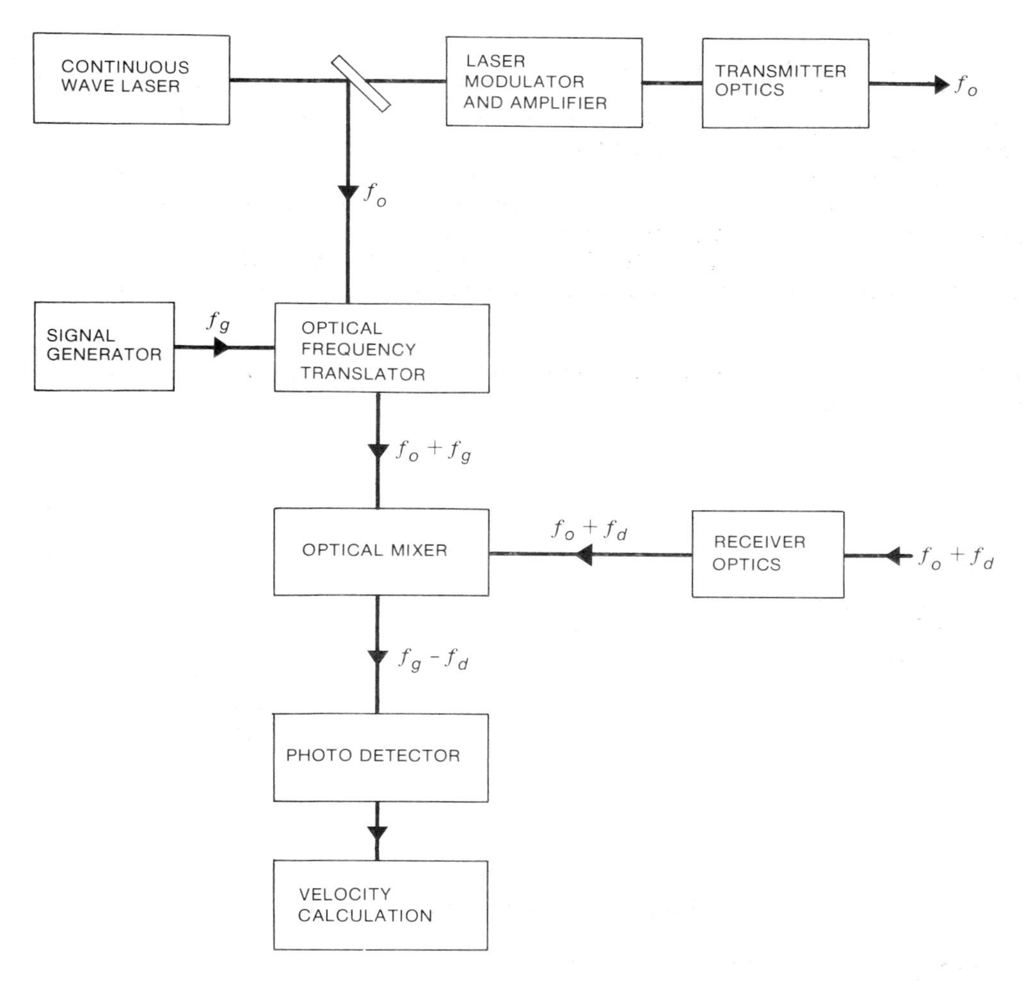

Fig. 5.14 Doppler system

The laser signal of frequency f_o is modulated and amplified before transmission to the moving target, which returns the energy shifted by the Doppler frequency f_d. As this frequency normally would be too high to be handled by the photodetector a reference frequency f_g is first combined with a fraction of the laser oscillator signal and the sum $|f_o + f_g|$ then mixed with the return signal to produce a frequency difference $|f_g - f_d|$ which is fed to the photodetector. Since f_g is known the Doppler frequency f_d is determined by measuring $|f_g - f_d|$. The radial velocity can then be deduced from the equation above and the target range obtained from the method employed in the pulse rangefinder.

The superior resolution of the laser tracker is offset by the narrow search cone, though this enables single targets to be selected. A mixture of laser and radar trackers offers the best compromise and some systems employ microwave modulation of the laser beam for increased sensitivity.

A high power CO_2 Doppler system at MIT works in modulated continuous working mode with an associated radar and visual tracker for acquisition. The device is used for tracking co-operative targets to a very high velocity accuracy.

The advent of laser weapons will impose severe requirements for high accuracy beam-aiming and tracking of fast moving and manoeuvring targets. A laser weapon is effective only if the laser beam can be aimed and maintained in a vulnerable point of the target until a kill has been achieved. For some targets this may require tracking jitter being held to better than one microradian in order to achieve good overlapping of the laser damage spot during the period of engagement. If an active tracker such as a subsidiary laser were to be used it would be important to choose one of the same wavelength as that of the weapon itself so that both will experience as close as is possible the same atmospheric transmission effects, and each should use the same boresight. Alternatively the high power laser weapon beam could be used for tracking as well as for inflicting damage by correlating the hot-spot which is produced with some reference point on the target using passive electro-optical techniques to record the images of these features.

Direct Fire Simulation

Lasers are used as a training aid for tank crews in precision gunnery against realistically moving targets, thus considerably easing the logistic problems, organisation, and ammunition costs associated with the conventional crew training exercises.

A typical laser based simulator operates in the following way. When the engaging crew is satisfied the weapon is laid on the target the gunner presses the firing button. A flash generator fires and the laser mounted coaxially with the tank gun emits pulses for one or two seconds. These are picked up by detectors on the target which automatically transmits a 'hit' or 'miss' signal by radio, the result of which is indicated in the gunner's eyepiece. If a hit has been scored a smoke flare is ignited on the target tank, its radio deactivated and its own laser switched off. These can be reactivated by an umpire or after a pre-set delay.

A tank gun training simulator is called SIMFIRE. It consists of a gallium arsenide laser emitting trains of pulses of 0. 6 mJ and 100 nsec width with repetition frequency ranging from 280 Hz to 300 Hz. The system is operational from 400 m to 2000 m.

The target detectors each cover 90° in azimuth and 35° in elevation; and four detectors are required to simulate a tank turret. The size of the target zone which will result in a 'kill' is 3 m x 2 m.

The radio link is crystal controlled and operated at 79 MHz. The link is activated only when responding to laser pulses received by one of the detector units. It then responds, pulse for pulse, with 2 μsec pulses of approximately 20 W average power.

A similar system built by Saab-Scania and called the BT41 Tank Combat Simulator is now in production for the Swedish Army. It features real-time simulation of the projectile flight, type of ammunition used and accurate simulation of target vulnerability characteristics. The system also generates a realistic tracer simulation into the gunner's sight. The laser beam is coded by modulation and is used to convey information to the target of the relative position between laser beam and projectile position, thus the target obtains data on the type and coordinates of the passing projectile and the identity of the attacking unit. The target has twelve detectors with retro-reflectors, thus two way information processing is possible. The relevant kill probability can be calculated from the coordinates of the projectile and the effect of the hit can be evaluated. The software enables varying conditional probabilities, eg chance of kill at second hit, to be evaluated. The BT41 can be used for training in direct fire gunnery for other targets, eg artillery and guided missiles, and target image simulators are built into the software.

FUTURE DEVELOPMENTS

Communications

Lasers offer great potential as a means of communication and are now being integrated into point to point and fibre optic communication systems. The narrowness of the laser beam offers high directional aiming and good protection against interference. They could be used for secret short range communication. Like its microwave counterpart the laser can be modulated to transmit either speech or pictures, by varying the amplitude of the beam with a shutter or by coding the pulses. The vast difference between them is the amount of information that can be carried by a laser beam. Even though the degree of modulation is lower with a laser beam the higher frequency enables bandwidths of the order 100 MHz of voice channel or 100 KHz of TV channel which are about one thousand times the microwave communication capacity. There are of course disadvantages, the main one being the atmosphere propagation loss.

Closed communication avoids the atmospheric transmission problem and is being examined using fibre optics. Low loss glass types have been developed and their potential for short range communication tasks has been established. They have two major advantages over electrical systems. Firstly their weight reduction of about one order of magnitude over conventional systems make them very attractive for lightweighting operations. Secondly they are free from interference from neighbouring electrical systems and from the electro magnetic pulse (EMP) from a nuclear explosion. Gallium arsenide has attractive qualities for use in fibre optic systems.

Holography

Laser holography is being considered for use in aircraft head-up displays to provide the pilot with a three dimensional view of the scene. The difference between a holographic image and an ordinary photographic image is that it has a stereoscopic appearance. The laser is used to take and record the holographic image which can be viewed by the pilot from different angles in much the same way as for direct viewing of the object.

Inertial Guidance

Lasers are now being used in inertial guidance systems. Their advantage over conventional gyroscopes are that they are smaller and weigh less, are cheaper to produce, are more stable, and take less time to make a reading. In operation two laser beams of the same wavelength travel in opposite directions around a square ring cavity. At one corner both beams are passed through a beam splitter and fed to photodiodes. The rate of rotation of the cavity is sensed by measuring the difference in frequencies between the two beams caused by one beam travelling further round the ring than the other, the difference between proportional to the rate of rotation. The difference can then be fed to servos which direct the aim point. The laser gyroscope has applications in aircraft and missile guidance and also as a very accurate angular measuring device for artillery fire control. Several types of laser are being used or developed for these purposes the most common one being the helium-neon laser which is built into the ring cavity. Nd:YAG and GaAs systems are also being developed, as are fibre optic types using an external laser source.

SELF TEST QUESTIONS

QUESTION 1 In what respects does laser light differ from natural light?

Answer ....

....

....

....

....

QUESTION 2 What are the main components of a laser?

Answer ....

....

....

....

....

QUESTION 3 What additional equipment is required if the laser is to be used as a pulsed rangefinder?

Answer ....

....

....

....

....

QUESTION 4 What are the main advantages of a laser over radar for rangefinding?

Answer ....

....

....

....

....

QUESTION 5 What are the main disadvantages of a laser over radar for rangefinding?

Answer ....

....

....

....

QUESTION 6 What would the size of a microwave transmitter radiating at 1 cm wavelength have to be to produce the same beam divergence as a carbon dioxide laser with a 5 cm diameter cavity?

Answer ....

QUESTION 7 Calculate the nominal ocular hazard distance for the naked eye against a single shot ruby laser rangefinder having the following properties:

Wavelength	0.694 μm
Pulse Energy	40 mJ
Pulse Duration	40 ns
Beam Divergence	0.5 mrad
Emergent beam diameter	5 cm

Answer ....

....

QUESTION 8 What would be the effect of using an eyefilter with 90% attenuation?

Answer ....

....

QUESTION 9 Calculate the nominal ocular hazard distance for the naked eye against a CW carbon dioxide laser having the following properties.

Wavelength	10.6 μm
Power	10 watts
Beam divergence	0.5 mrad
Emergent beam diameter	5 cm

Answer ....

....

QUESTION 10 What precautions must be taken if the lasers defined in questions 7 and 9 are to be used in the field against static targets?

Answer ....

....

....

....

....

....

....

....

....

....

ANSWERS ON PAGE 192

6.

Radar

INTRODUCTION

The basic principles are outlined in a section of Chapter 1 and the reader is recommended to read this first. To sum up, radar measures:

Range - by timing the passage of electro magnetic waves to and from a target. To do this it is necessary to label the wave in some way; eg by sending out a short pulse of waves.

Direction - by forming the waves into a narrow beam by a suitable antenna, so that the direction of the antenna gives the direction of the target.

Relative velocity or target movement - by the Doppler effect.

The roles of radar in surveillance and target acquisition include the detection and location of aircraft, of moving targets on the battlefield, of ships and even of projectiles, such as rockets and shells. Among peaceful applications are the control of aircraft, the detection of rain storms, maritime navigation aids and intruder alarms. Radars can be static, they can be mounted in vehicles, ships or aircraft, they can be man portable or simply hand-held. They range in size from enormous permanent structures for long range air surveillance to pocket-size for indicating movement at a distance of a few metres.

In Chapter 1 it states that radar gives a ready indication of range: it can penetrate dust, fog etc. and it can be used at night. On the other hand it is often bulky with a prominent antenna; it is active, leaving it open to countermeasures, and its angular resolution and its ability to recognise targets are poor.

RANGE MEASUREMENT

Perhaps the best known feature of radar is its ability to measure range accurately. Historically the oldest method of range measurement is to send a pulse of radio frequency waves and to time the return of the echo from a target; this is still by far the most usual method. The principles are shown in Fig. 6.1.

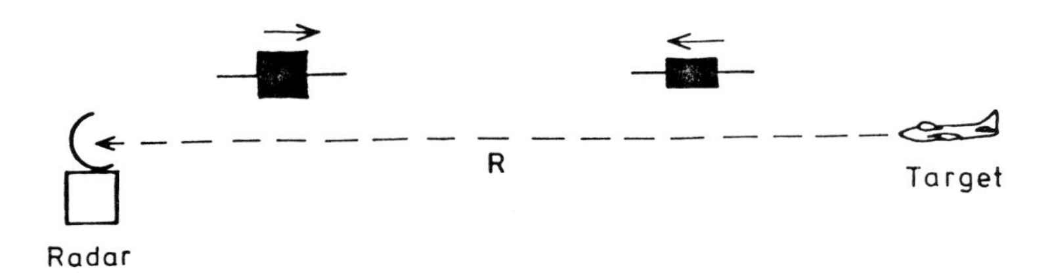

Fig. 6.1 Pulse range measurement

R is given by $T = 2R/c$, where T is the elapsed time between transmitting the pulse and receiving the echo; c, the velocity of propagation of radio waves, is 3×10^8 m/s. T is very short as this example shows:-

$$R = 15 \text{ km} \therefore T = 2 \times 1.5 \times 10^4 \div 3 \times 10^8$$
$$= 10^{-4} \text{ seconds or } 100 \text{ μs.}$$

For such short time intervals it is necessary to use electronic timing. Older radars use an analogue method with a Cathode Ray Oscilloscope (CRO) (Fig. 6.2 (a)).

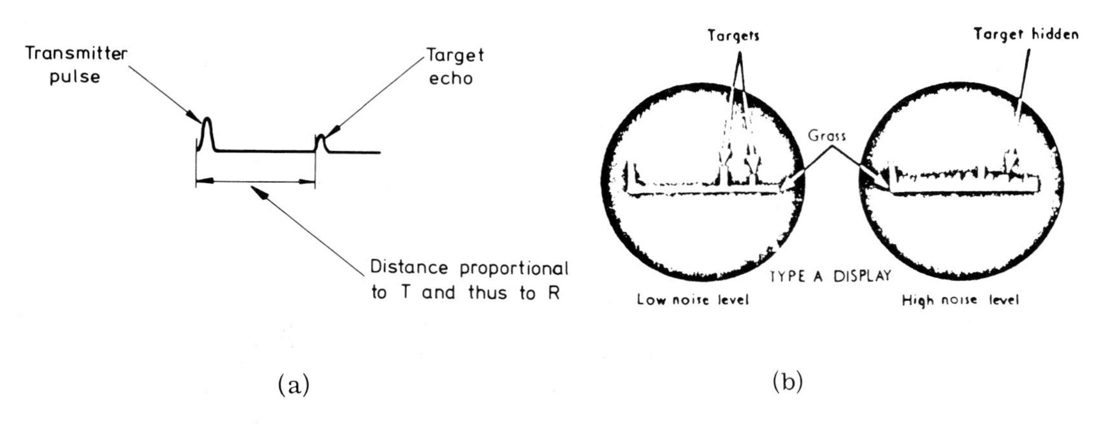

Fig. 6.2 CRO method of range measurement

Provided the time-base of the CRO is linear the distance of the echo from the beginning of the time-base display on the tube face is proportional to R. (b) shows a typical display. In modern equipments the tendency is to use digital methods of time measurement as in the well known quartz crystal controlled watches and clocks (Fig. 6.3).

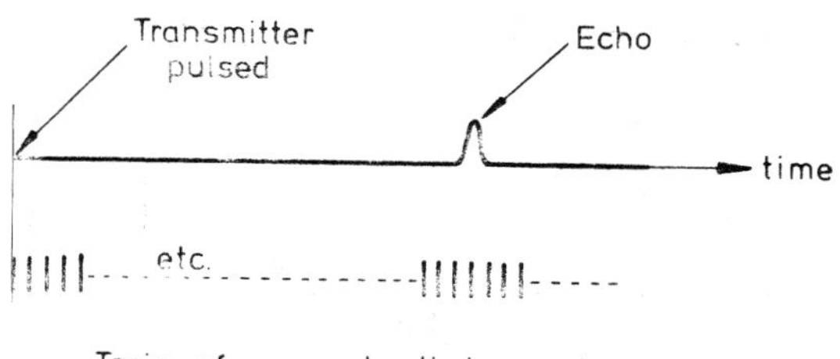

Fig. 6.3 Digital range measurement

Counting the number of timing pulses from the instant of transmission to the return of the echo gives the elapsed time to a high degree of accuracy (1 part in 10^6 or better).

The radar does not usually transmit just one pulse but sends out a train of pulses at a fairly high pulse recurrence frequency (p.r.f.) - typically 1 or more kHz. A high rate is necessary to avoid flicker on a CRO display and to provide adequate confirmation of the presence of a target; the radar uses the highest p.r.f. practicable. The minimum time between transmitted pulses is the time for an echo to return from a target at the maximum operational range of the radar and this sets the upper limit on the p.r.f. For example if the maximum operational range is 75 km, then the elapsed time is 0.5 ms, and the p.r.f. $\leq$ 2 kHz. In practice it is necessary to allow some dead time so that the radar shall ignore echoes from exceptionally large targets at beyond the operational range; in this example a p.r.f. of 1.5 kHz might be a suitable choice, giving a dead time of 167 μs.

Radar range measurement systems are capable of a high degree of accuracy (1 part in 10^6 has been mentioned); to take advantage of this the pulse duration should be short. If a pulse has a duration τ seconds the difference in elapsed time between the beginning and the end of an echo from a 'point' target is also τ seconds, hence the echo spreads over a range interval of $c\tau/2$ metres. This spread is known as the range resolution ΔR. and should be comparable with the accuracy of the timing in the radar. Typical pulse durations lie in the bracket 0-1 to 2 μs; corresponding to ΔR of 15 m to 300 m.

Pulsing is not the only form of labelling or modulation of the transmitted wave; in fact, any form of modulation will do, though practical considerations limit the number of types in use. A common alternative to pulse is linear Frequency Modulated Continuous Wave transmission (FMCW). (See Fig. 6.4).

The instantaneous difference in frequency between the transmitted and received signals gives the elapsed time T, and hence R. The time resolution is $1/\Delta f$, where Δf is the total frequency deviation of the transmitted signal, ie Δf of 10 MHz gives the same time, and range, resolution as a pulse of 0.1 μs in a pulse system. The advantage of FMCW is that the transmitter operates at a constant power level, a more favourable operating condition than generating the same average power but in a train of short high powered pulses. The disadvantages are possible confusion of range measurement with Doppler shift if the target is moving, and difficulties of using one antenna for both transmission and reception.

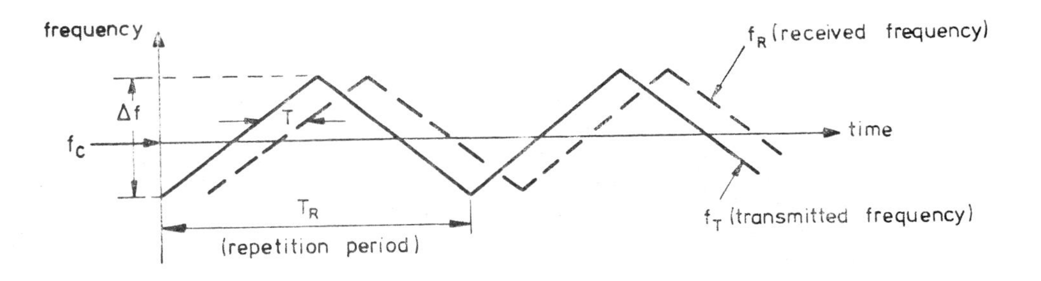

Fig. 6.4 FMCW range measurement

DIRECTION FINDING

The antenna concentrates the radar power into a very narrow beam, often only a few mils wide; only targets within the beam return echoes, hence the direction of the source of an echo is known to a high degree of accuracy.

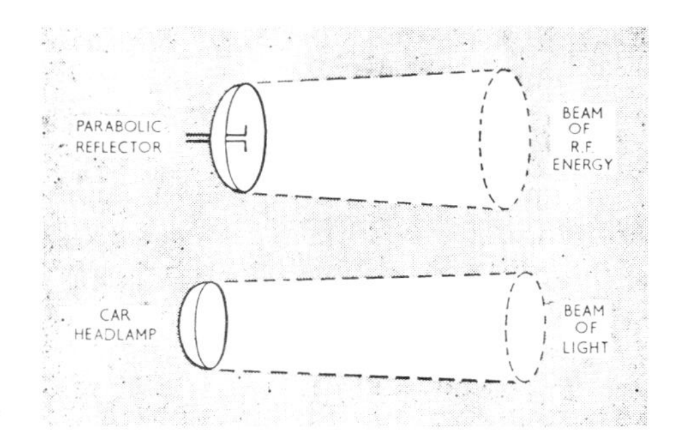

Fig. 6.5 Direction finding by narrow beam

However, to obtain such a narrow beamwidth, the width or diameter D of the antenna must be large in comparison with the wavelength. A useful rule of thumb is that beamwidth in mils is between 1000 and 1500 times λ/D. Eg, for a 10 mils beamwidth, D must be at least 100 λ wide. It follows that λ must be short for the antenna to be of practicable size; eg, for D not to exceed 3 metres, λ must be no longer than 30 mm, corresponding to a frequency of 10 GHz. In the early days of radar it was not possible to generate useful powers at such high frequencies, much lower frequencies were used so that the angular accuracy was poor. It was not until the invention of a really high powered 'microwave' source, the cavity magnetron, in 1940 that modern radar angular precision became possible. Modern radar equipment frequencies are concentrated in the microwave and millimetre wave bands (ie wavelengths measured in cm or mm) as shown at Annex A. Note the systems of classification by letter band as well as by frequency or wavelength; eg in the NATO classification scheme, a frequency of 10 GHz is in 'I Band'.

A TYPICAL SURVEILLANCE RADAR

As an example of the techniques of target detection and location, consider a simple pulse radar for all round air defence. The radar is to give location in range and azimuth only, but is to cover all targets up to a given height. The antenna is broad, so as to give a narrow beam in azimuth, and it rotates at a constant rate for all-round surveillance. The beam is wide in elevation to give the required height coverage (fan beam). Figure 6.6 shows the idea, together with typical antenna shapes.

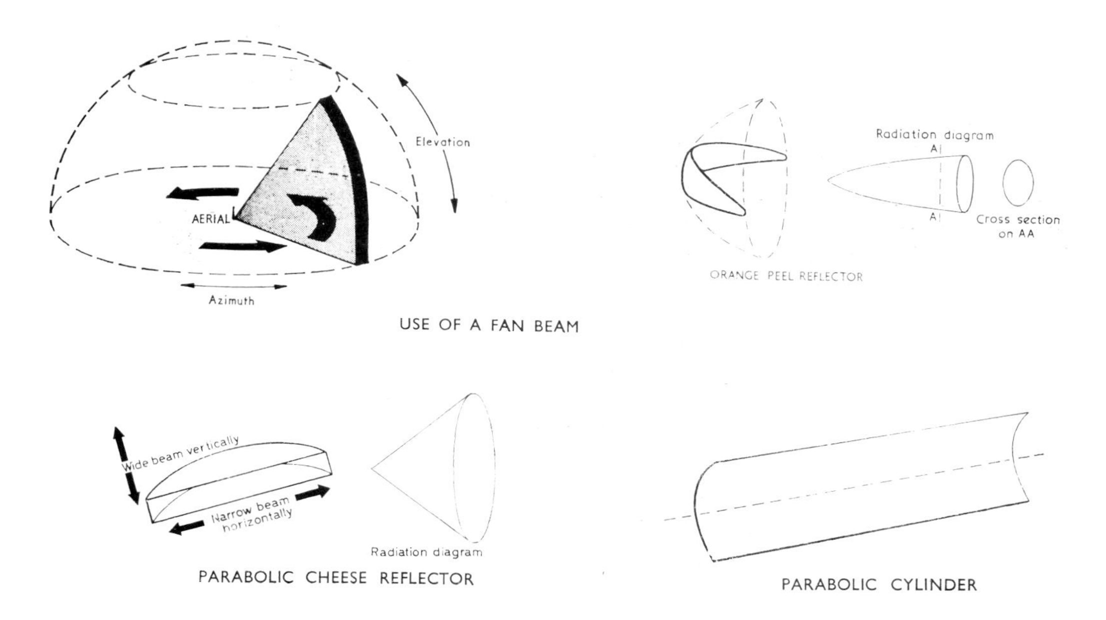

Fig. 6.6 Producing a fan beam

The antenna rotates at several rpm; the rotation speed must be sufficiently high to ensure that targets are detected early enough and that subsequently the information is up to date. For example if the rotation rate were 6 rpm an aircraft travelling at 660 m/s could penetrate 6.6 km into the surveillance volume before the radar detected it. On the other hand the higher the scan rate the greater the air resistance and the more the turning power; the turning rate is therefore no higher than necessary.

Although a simple fan beam in elevation of adequate width can give the required height coverage, it is wasteful of transmitted power. Air defence vertical coverage is usually of the shape shown in Fig. 6.7 (a) specifying a maximum range at low angles of elevation, and a maximum height coverage up to a certain angle of elevation. From Fig. 6.7 (b), which shows an idealised shape, the antenna generates a fanbeam covering E_1 with maximum radiation.

(a)

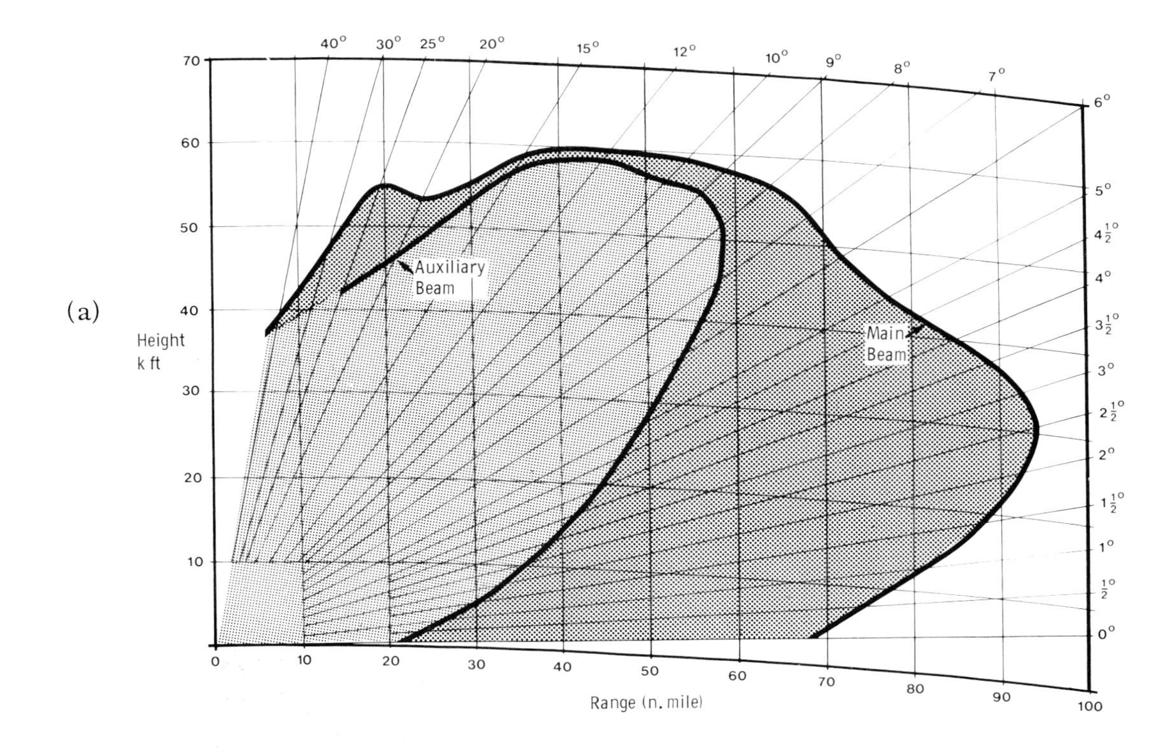

(b)

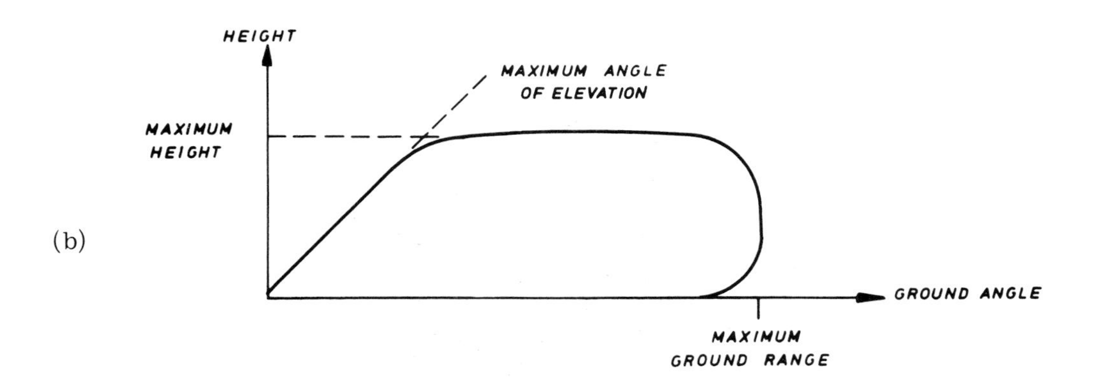

Fig. 6.7 Air defence, vertical coverage diagram

In the upper region, E_2-E_1, the higher the elevation the less the range, hence the antenna covers this region with radiated power which decreases progressively with angle of elevation and which cuts off sharply at E_2. To achieve this the reflector, which would otherwise be parabolic, is curved more sharply at the bottom to divert power into the upper region according to the requirement of the coverage diagram. Figure 6.8 shows an antenna of this type and the associated beam shape, known as cosecant^2 because of its geometry.

Fig. 6.8 Cosecant^2 antenna

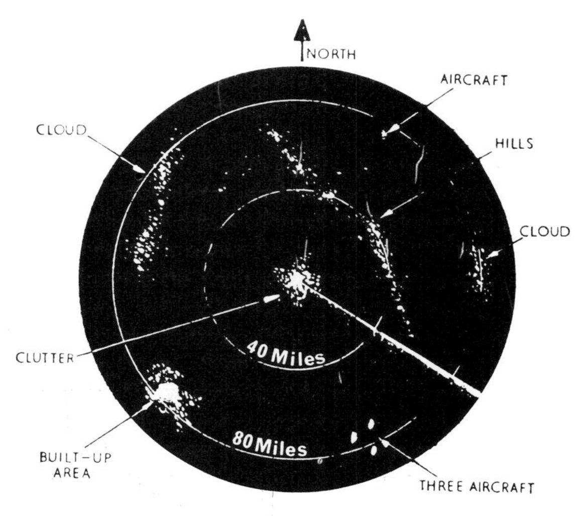

Fig. 6.9 PPI display

The operator's display (b) is a type of CRO known as a Plan Position Indicator (PPI). The range time-base extends from the centre of the tube face to the edge so that radial distance of an echo represents range. The time-base rotates in synchronism with the antenna so that the direction of an echo from the centre of the tube face gives the azimuth of the target. Target echoes appear as a brightening of the time-base; the tube screen has a long persistance so that the bright spot of target echoes remain on the screen, lasting for at least one complete revolution of the antenna. The operator can read off range from a cursor pivoted over the face of the tube, and azimuth from a scale around the perimeter. More usually he controls electronic markers which he places over the target echo, and in doing so he generates electrical signals which can indicate target data to a distant centre etc. Figure 6.10 shows the block diagram of the radar.

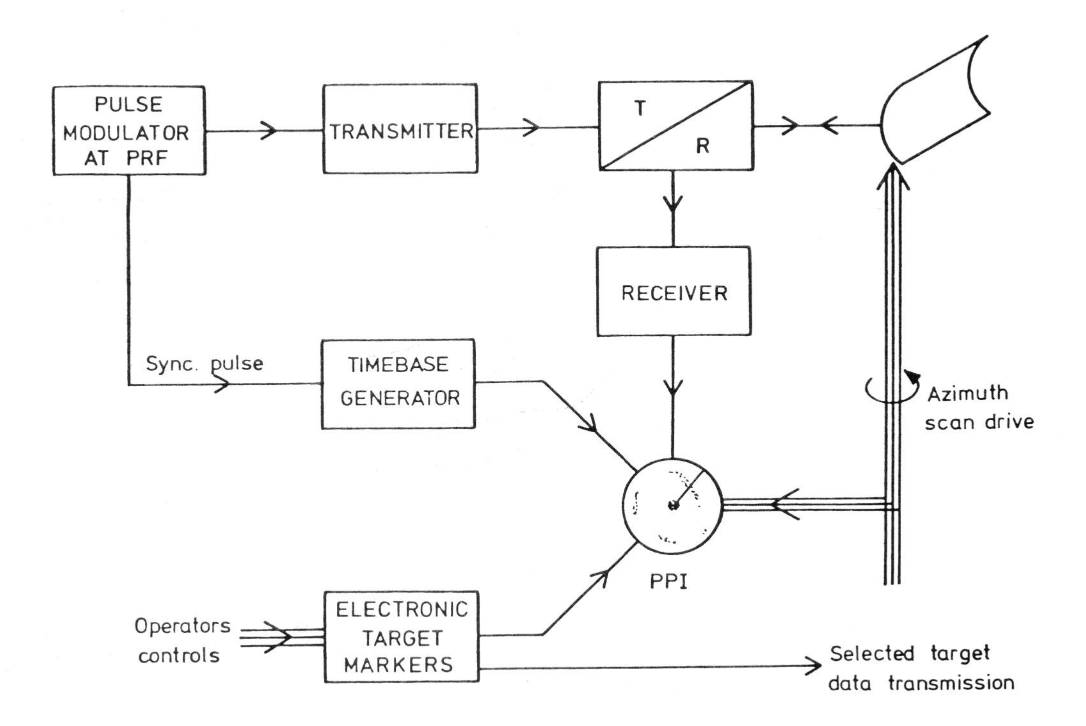

Fig. 6.10 Air defence surveillance radar - block diagram

Note the sync pulse from the pulse modulator to the time-base generator to start the timing. Note also the Transmit/Receive switch (T/R) so that one antenna can serve both purposes. The switch normally connects the antenna to the receiver; the start of the transmitter pulse immediately disconnects the receiver and connects the transmitter to the antenna; at the end of the transmitter pulse the switch relaxes automatically to reconnect the receiver.

The fan beam or the cosecant^2 beam give elevation coverage but no measure of angle of elevation of a target. To obtain this it is necessary either to use a separate height finding radar or to incorporate elevation measurement into the antenna system of the surveillance radar. A separate height finding radar has an antenna beam which is narrow in elevation, though wide in azimuth, a beaver-tail beam, Fig. 6.11 (a). The surveillance radar indicates the azimuth of the target; the height finder Fig. 6.11 (b) slews to this azimuth and the antenna nods, scanning a sector in elevation.

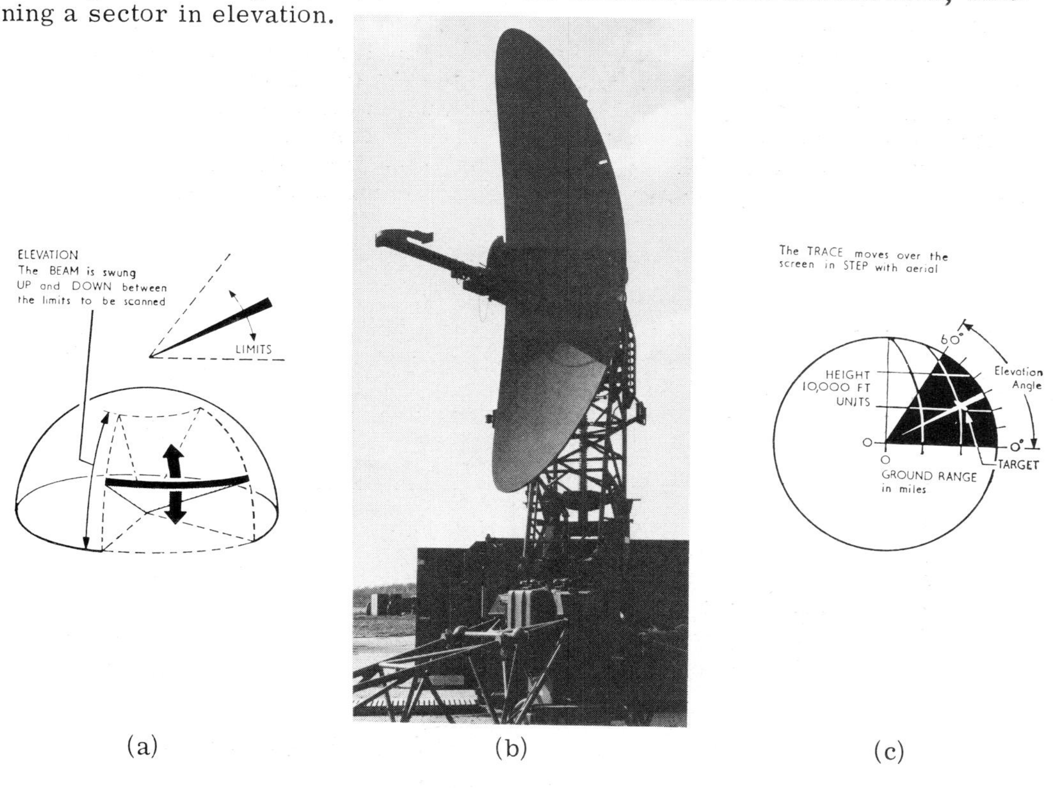

(a) (b) (c)

Fig. 6.11 Height finding radar

Target echoes appear on a Range Height Indicator (RHI) (Fig. 6.11 (c)) which is similar to a PPI but covering a sector only. The operator identifies the target by its range and reads off the elevation and height.

If the surveillance radar incorporates its own height finding the elevation coverage is split up into a number of narrow sectors (Fig. 6.12). Either multiple antenna feeds or a special beam forming matrix generate the multiple beams necessary. This leads to quite a complicated radar because each beam requires its own receiver. Multiple displays might be possible but it is more usual to pass the target data to a computer for storage and evaluation. Sometimes selected targets are presented on a synthetic PPI display from the computer store, height being indicated by figures alongside the target echo (Fig. 6.13). Because it measures all 3 coordinates this type of radar is sometimes known as 3-D.

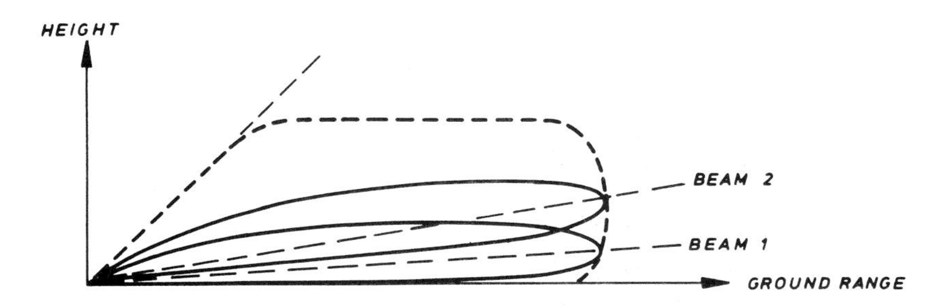

Fig. 6.12 Multiple beams in elevation

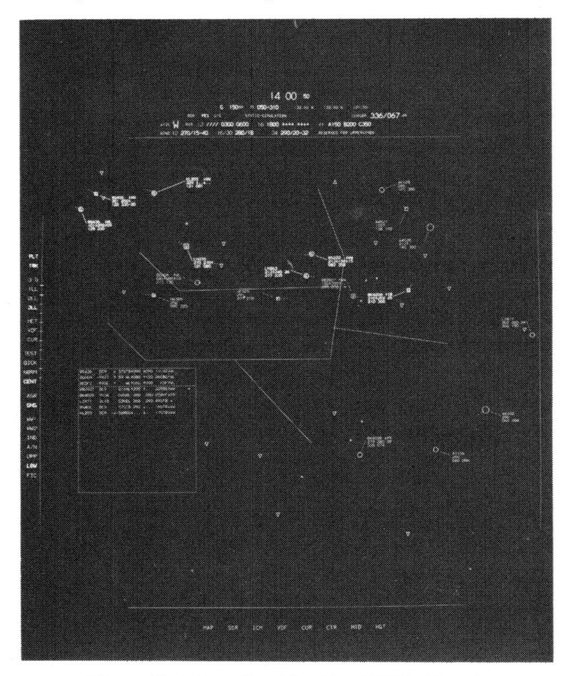

Fig. 6.13 Synthetic PPI display

BATTLEFIELD AND SIMILAR SURVEILLANCE RADARS

Battlefield surveillance radars, projectile detection radars and some types of air surveillance radars cover only a limited sector in azimuth. Battlefield surveillance radars and other types of small ground surveillance radars have quite a small antenna which an operator turns by manual control; indeed, the radar may be so small that 'scanning' consists merely of turning the whole radar. The antenna is often a simple paraboloid mirror, but a modern technique is to use a lightweight array on a thin rectangular sheet of plastic, as shown in Fig. 6.14. Ranging in these radars is usually digital with only a numerical display.

Fig 6.14 Lightweight array antenna for small radars

Since only moving targets are of interest these radars use the Doppler effect to exclude stationary echoes from the ground (clutter) and to give an estimate of the velocity of the target under investigation.

Projectile detection radars for locating mortars and artillery scan the sector much more rapidly, several times per second, to be sure of intercepting any projectile in the sector at least once while it is still low down on its trajectory. The fast scan rules out movement of the whole antenna, instead radars have a stationary main reflector while rapid movement of some part of the internal feed system, as in the 'Foster Scanner', causes the beam to scan. Figure 6.15 is a diagram of the principle.

Projectiles appear as bright echoes on a cartesian coordinate cro display giving azimuth and range, while the known fixed elevation of the beam gives projectile elevation. The latest weapon locating radars, particularly those for artillery location, use inertialess electronic scanning, described later in the chapter, combined with automatic target detection and data processing.

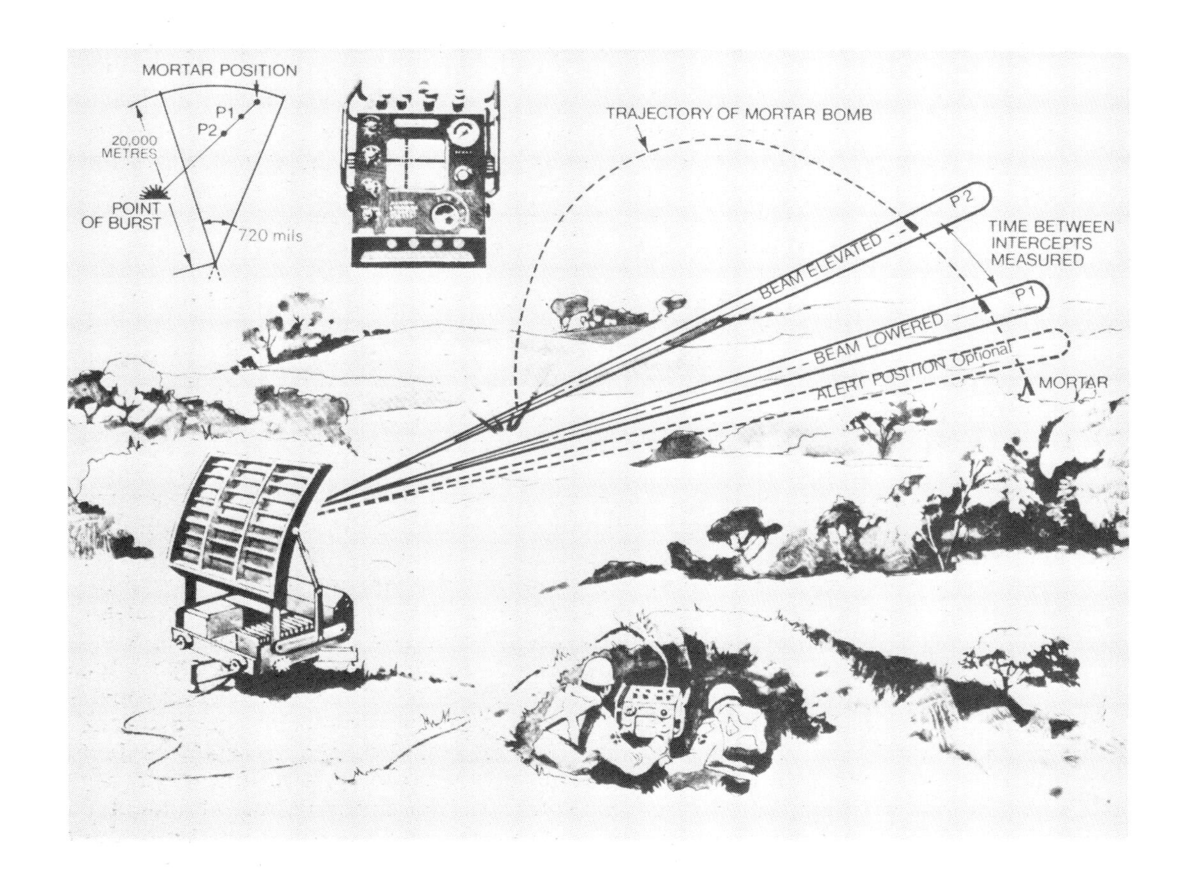

Fig. 6.15 Mortar location radar

SYSTEM PERFORMANCE

One of the most important parts of the specification of a radar covers its ability to detect a target at an adequate range. This section studies the factors which govern this ability.

The first of these factors is the ability of the antenna to concentrate power into a narrow beam and, conversely, to receive signals only within the same narrow beamwidth. This concentrating ability is known as the Power Gain, G. Consider an antenna which concentrates its radiated power into a solid angle Ω steradians, instead of radiating uniformly in all directions, ie over 4π steradians. The concentration, or G, is $4\pi/\Omega$; in radar antennas G can be many thousands. The concentrating ability of an antenna applies also to receiving properties; it extracts power from an incident wavefront over an area A, the effective aperture area, which is related to G by $A = G\lambda^2/4\pi$. In quasi-optical antennas, such as parabolic mirrors and in large planar arrays, A is nearly equal to the physical area of the antenna aperture. For example a circular paraboloidal antenna of diameter 1 m has A about 60% of $\pi/4 m^2$, and a gain at $\lambda = 30$ mm of about 7000.

Unlike natural electro magnetic radiation such as light, radio waves are usually 'polarised' in a single plane. The plane of polarisation is defined as that which contains the electric field vector. The orientation of the feed of the transmitting antenna determines the plane of polarisation and the receiving antenna must accept this same plane of polarisation.

The ability of a radar to detect a target depends obviously upon the nature of the target. The relevant properties are the size, the material from which it is constructed (metal reflects better than non-conducting materials), and its aspect: a flat surface presented normally to the radar reflects much better than one which is inclined; rounded surfaces and points also reflect poorly. The strength of a target echo depends also upon λ and upon the plane of polarisation.

Practical targets consist of many reflecting surfaces, some of which predominate, such as the flat sides of a tank, the wings of an aircraft and fins on a projectile; the net echo is the resultant of these individual reflections. Because the individual reflections depend upon aspect, the net echo also depends upon it as can be seen from the 'polar diagram' of a target at Annex B; hence it is usually necessary to specify aspect when defining the echoing properties of a target. In passing, it should be noticed that the rapid changes of echo strength with aspect lead to troublesome fading. To gain an idea of the definition of these echoing properties, consider a metal sphere of radius a; the sphere intercepts the power from an area of incident wave equal to the projected area of the sphere, πa^2; neglecting any loss in the metal, all of the intercepted power is re-radiated, and because a sphere is symmetrical, it re-radiates isotropically, acting as an antenna of power gain 1. Most practical targets are not symmetrical, and do not re-radiate isotropically; nevertheless one can consider the observed or calculated re-radiation towards the radar receiving antenna to have come from an equivalent metal sphere; whose projected area is known as the radar echoing area, or radar cross-section, σ of the target. σ of a meteorological sphere of diameter 1 foot is simply 0.28 m^2; typical values of σ for more complicated targets are in the following table:-

TABLE 1 Typical Values of σ

Serial	Target	(m^2)
1	Ship	1000s
2	Large aircraft plan view	500
3	Large missile plan view	100
4	Large aircraft slant view	10
5	Small aircraft slant view	2 - 5
6	Tank	1 - 10
7	Missile	.03
8	Bomb	.005
9	Shell	.0005

The power incident upon a target from the transmitter is inversely proportional to the square of the range, ie to R^{-2}; similarly, assuming the transmitting and receiving antennas are co-located, the re-radiated power received from the target is also proportional to R^{-2}; hence the received signal power or strength is proportional to R^{-4}. If the transmitter and receiver are in different places then the received power is proportional to $R_1^{-2} x R_2^{-2}$; where R_1 and R_2 are the 2 ranges concerned.

The received signal power S has to be observed against an unavoidable background of noise. Even if the receiver were completely noise free the minimum noise power would be equal to kT_0B, where k is Boltzmann's constant, T_0 is the ambient temperature in K and B is the bandwidth of the receiver. At $T_0 = 290$K, kT_0 is conveniently 4×10^{-21} watts per Hz. However practical receivers add their own quota of noise, degrading the signal: noise by a factor known as the Noise Figure, F. F is typically 10-30, though lower values are possible with special receivers.

How does the receiver bandwidth B enter into the considerations? A pulse of radar frequency waves has a broad spectrum of frequencies and the bandwidth must be wide enough to accommodate this; on the other hand it must not be so wide that it allows the passage of too much noise. In practice the bandwidth which maximises the S:N for a pulse of duration τ is $1/\tau$ Hz approximately.

Another factor which determines S:N is the time t_0 the target is under observation by the radar. Ideally S:N is proportional to t_0; in practice the proportionality lies between t_0 and $t_0^{\frac{1}{2}}$. One usually accounts for this by multiplying t_0 by an appropriate Integration Efficiency Factor, E_i, which is less than 1. In a surveillance radar t_0 is the time the target lies in the beamwidth; eg antenna scan rate 10 rpm, beamwidth 20 mils, t_0 on a point target is about 20 ms.

Losses arise from various causes, attenuation in feeders and in the T/R switch; degradation in performance during service life etc. An important loss at higher frequencies is that due to attenuation in the atmosphere, by absorption of the signal by oxygen and water molecules. The form of the attenuation is shown in Fig. 6.16. It is negligible at $\lambda = 22$ mm; note the peaks and the troughs, or windows, at shorter wavelengths; the windows at 8 and 3.5 mm are useful for short range

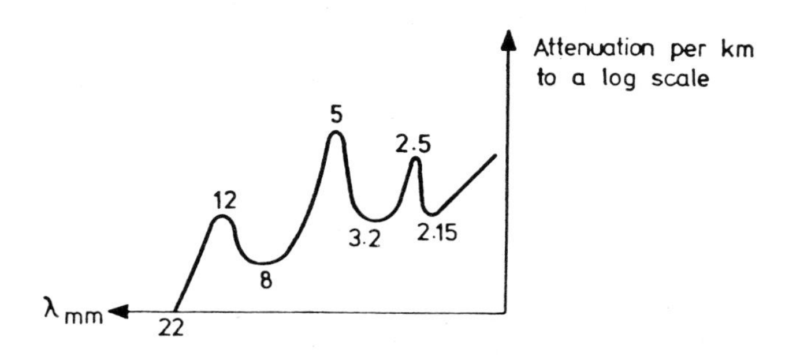

Fig. 6.16 Atmospheric attenuation

mm radar. Attenuation by rain is more severe and is significant at wavelengths as long as 0.1 m. Detailed numerical graphs of both atmospheric and rain attenuation are available in text books on radar and on radio wave propagation.

We are now in a position to write down an equation relating receiver signal:noise to transmitter power and target range. A general form of this equation for monostatic radar is:-

$$S:N = \frac{P(\text{average})\ t_0 E_i\ GA\sigma}{(4\pi)^2 LF\ kT_O\ (B\tau)R^4}$$

Assuming the receiver bandwidth B is matched to τ, so that $B\tau = 1$, and that a common antenna is used for transmission and reception, so that $A = G\lambda^2/(4\pi)$, the equation becomes:-

$$S:N = \frac{P(\text{average})\ t_0 E_i\ G^2 \sigma\ \lambda^2}{(4\pi)^3\ LF\ kT_O R^4}$$

Note that it is the average power which counts not the pulse power. As an example of the use of this equation and to give an idea of the magnitudes of the quantities, calculate P(average) for a radar to detect a target of $\sigma = 1\ m^2$ at R=40 km; G=2000, t_0=2.0 ms, E_i=0.6, L=10, F=10 and λ =0.1 m. To detect a target requires S:N at the output of 50:1.

$$\therefore\ P = \frac{50 \times (4\pi)^3 \times 10 \times 10 \times 4 \times 10^{-21} \times (4 \times 10^4)^4}{2 \times 10^{-2} \times 0.6 \times 2000^2 \times 1 \times 0.1^2}$$

$$= 212 \text{ watts (average)}$$

If the p.r.f. were 1 kHz and the pulse duration 1 μs the power in each transmitted pulse would be $212/(10^3 \times 10^{-6}) = 212$ kW. These powers are well within the capability of available transmitter valves.

OTHER FACTORS AFFECTING PERFORMANCE

Unlike the longer wavelengths, radio waves used in broadcasting etc., microwaves cannot propagate for any appreciable distance over the horizon. Provided h_1 and h_2 (Fig. 6.17 (a)) are $\ll$ a, as they will be, the horizon distance is:-

$$R_H = \sqrt{2ah_1} + \sqrt{2ah_2}$$

In fact microwaves propagate slightly over the optical horizon because the atmosphere bends radio waves downwards (Fig. 6.17 (b)). This is because the air density falls with height so that the lower edge of the wavefront travels more

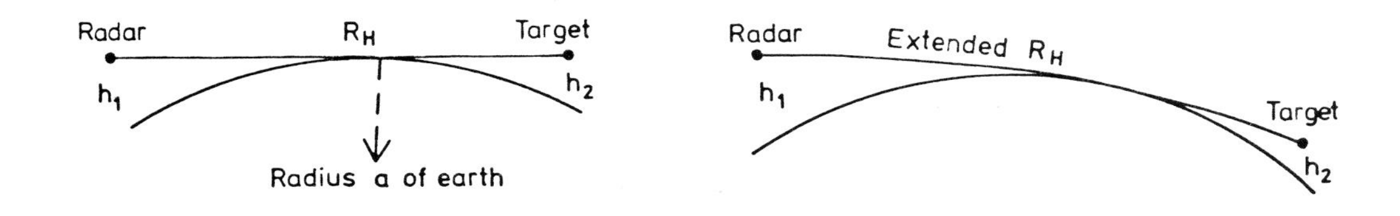

Fig. 6.17 Horizon

slowly than the upper part which is in a less dense medium. Applying a correction for this under standard atmospheric conditions extends R_H by about 15%. A useful formula for this radar horizon is:- 4130 ($\sqrt{h_1}$ + $\sqrt{h_2}$) metres where h_1 and h_2 are in metres. The following example shows that this limits the detection range of ground radar on low flying targets; h_1 = 3 m, h_2 = 15 m, giving R_H = 23 km. Over hilly terrain the horizon may well be much less, due to screening.

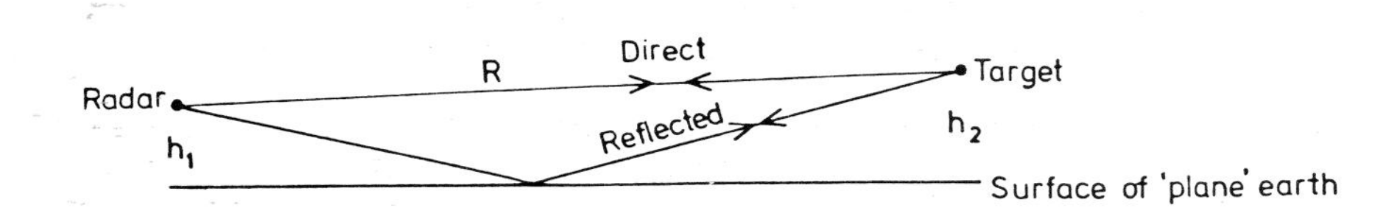

Fig. 6.18 Plane earth reflection

Another factor affecting the ability of a radar to detect low flying targets is interference between the direct path and a path reflected from the surface of the earth. If h_1 and h_2 are small compared with R (Fig. 6.18), so that the grazing angle for reflection is small, almost complete interference takes place between the direct and reflected waves; this effect is particularly marked over the sea, whose conductivity is high. The effect of this interference is that the signal strength falls rapidly at ranges beyond about $12h_1h_2/\lambda$. Using the heights of the previous example and taking λ = 0.1 m shows that this rapid fall of signal strength takes place at beyond 5.4 km: this is obviously a severe limitation and to alleviate it entails either raising the antenna height or decreasing λ . (Note, the theory assumes a plane earth, a valid assumption provided $R \ll R_H$; the curvature of the earth exacerbates the effect).

Occasionally atmospheric conditions near the surface of the earth lead to greatly extended ranges at low angles of elevation. If the air density just above the surface of the earth falls much more rapidly than usual the radar waves may be bent downwards so much that they follow the curvature of the earth giving rise to a duct in the coverage diagram, Fig. 6.19. Ducts are only a few metres high hence they can only trap short wavelengths such as in radar. They occur particularly in hot climates over the sea, and over hot deserts at night and can give anomalous radar ranges of several hundred miles.

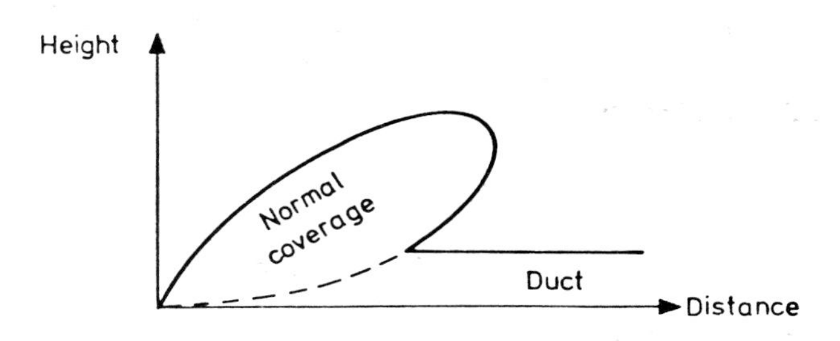

Fig. 6.19 Atmospheric ducting

DOPPLER RADAR

Radar uses the Doppler effect to detect movement of a target relative to the radar and, if necessary, to measure the relative velocity. If a target is moving relative to a radar with velocity u the received signal frequency differs from the transmitted frequency by the Doppler frequency, given by $f_D = u\frac{2f_c}{c}$ or $\frac{2u}{\lambda}$; f_D is positive for the approaching target, negative for a receding one. Note that it is only the relative velocity which causes the Doppler effect. A vehicle moving towards a radar at 30 mph (13.4 m sec^{-1}) gives a Doppler frequency of about 900 Hz with a radar $\lambda = 30$ mm, a frequency well within the audible spectrum. On the other hand an aircraft flying at 660 m sec^{-1} gives a Doppler shift of 44 kHz at the same λ.

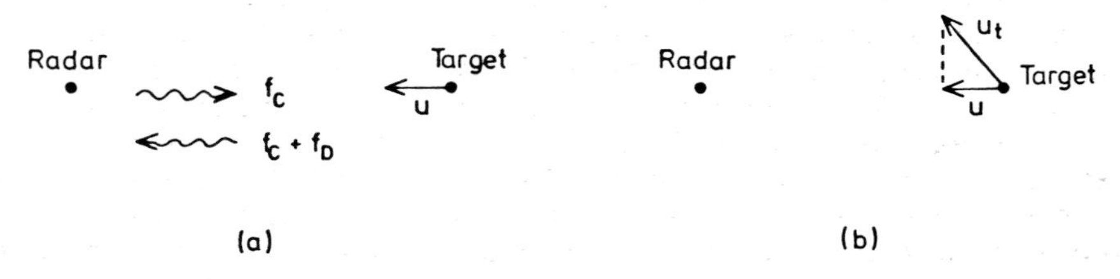

Fig. 6.20 The Doppler effect

Figure 6.21 shows the block diagram of a simple cw radar for detecting intruders. The radar compares transmitted and received signal frequencies; a difference indicates the presence of a moving object, and can sound an alarm, while measuring the Doppler frequency gives target velocity as in police traffic radars etc. The purpose of the clutter filter is to remove signals with zero or nearly zero f_D; ie signals from stationary objects. The circulator enables the radar to use the same antenna for transmission and reception.

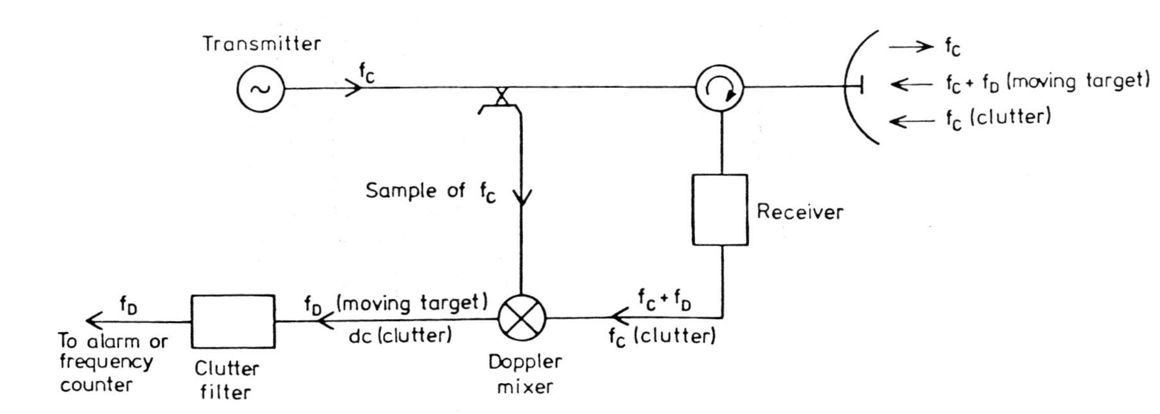

Fig. 6.21 Doppler radar

Ordinary pulse radars for the detection of low flying airborne targets or surface targets suffer badly from clutter, which can obscure completely small moving targets. Modifying the radar so that it detects Doppler shift enables it to see the small moving targets through a strong clutter background. Such a radar is similar to the cw radar of Fig. 6.22, except that the transmission is pulsed in an amplifier following the cw oscillator. The radar uses conventional range measurement but a clutter filter rejects echoes from stationary targets.

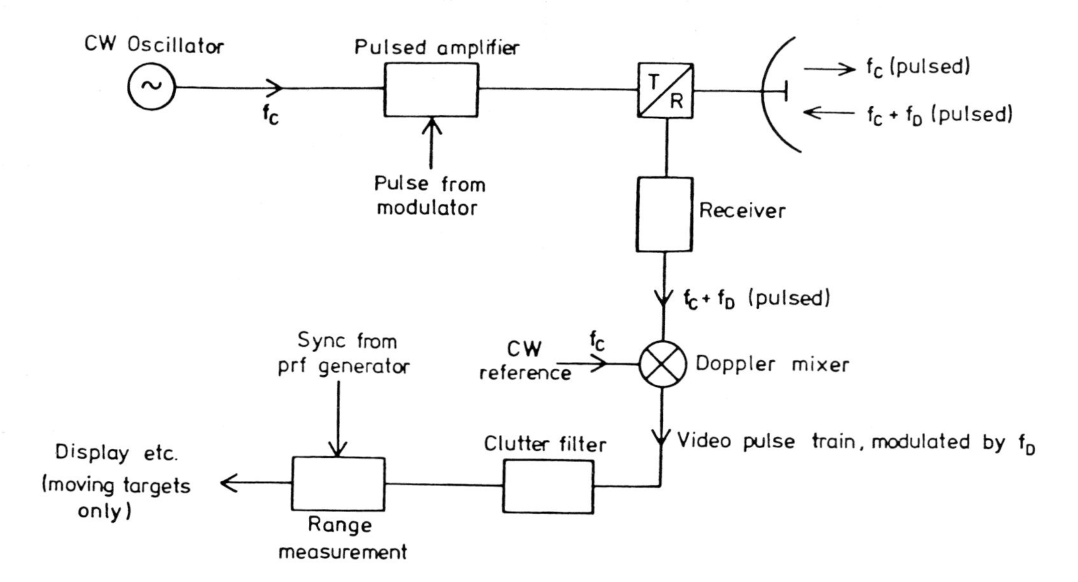

Fig. 6.22 Pulse radar with Doppler MTI

Figure 6.23 (a) shows the PPI display of a conventional pulse air surveillance radar; note the heavy clutter. Figure 6.23 (b) shows the display of the same radar but with this Moving Target Indication (MTI) facility. Most modern air defence surveillance radars now incorporate MTI.

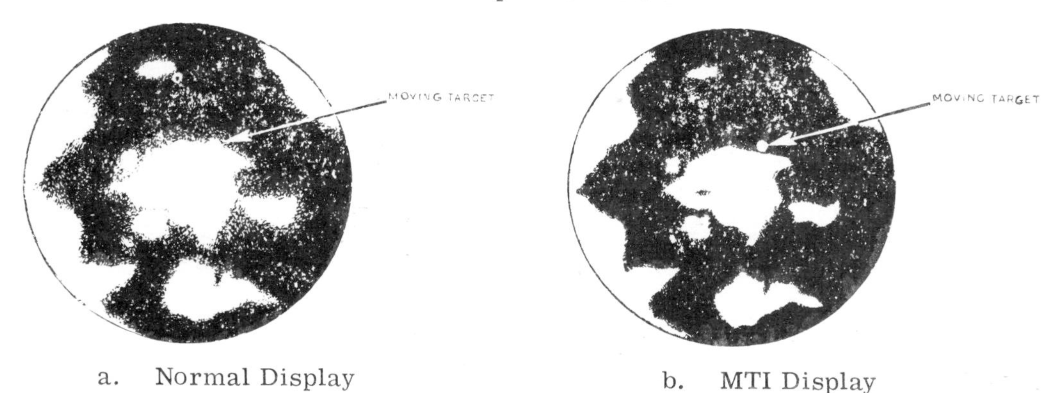

a. Normal Display b. MTI Display

Fig. 6.23 PPI with MTI

The same type of radar can actually measure the velocity of every moving target by measuring its Doppler frequency, or frequencies, after the range measurement. A battlefield surveillance radar would require no more than a pair of earphones for the operator to hear the note of the Doppler frequency; on the other hand air defence radar would require an elaborate system of frequency filters to segregate Doppler frequencies from many targets which might be present in a massed raid. To assess the mass of target data from such a radar would be beyond the capabilities of an operator; data storage and assessment would be a task for a computer. Although there is no basic difference between this type of radar and the MTI radar, pulse radars which use the Doppler effect actually to measure velocity are generally known as 'Pulse-Doppler' radars.

One difficulty arises in radars which both measure range and use the Doppler effect. If the two frequencies are equal, or if the Doppler frequency is a multiple of the p.r.f., the two frequencies interfere in such a way that the Doppler frequency appears to be zero; ie as if the target were stationary. The clutter filter then rejects the target echo and the radar is 'blind' to targets of the corresponding speeds. For example an air defence surveillance radar with MTI can expect targets flying at speeds up to 600 m sec^{-1}, the p.r.f. is 2 kHz and $\lambda = 0.1$ m. $\therefore$ f_d lies in the range 0-12 kHz; within this frequency range there are no less than 6 blind speeds - every 100 m sec^{-1}. On the other hand a battlefield surveillance radar would not expect to encounter a target travelling faster than about 30 m sec^{-1}, the p.r.f. would probably be higher and there would be no difficulty with blind speeds. Where blind speeds do exist, a palliative is to use the longest wavelength possible; however there are limits to this and the usual solution to the problem is to use several different p.r.f.s, so that the radar is never totally blind in the range of target speeds of interest.

SURVEILLANCE FROM THE AIR

Aircraft, drones and satellites are particularly suitable for surveillance of the ground; they are also valuable for surveillance of airspace because of the long distance to the horizon. An example of the former is Sideways Looking Airborne Radar (SLAR), and of the latter, systems like AWACS.

SLAR has a long linear antenna in a pod carried longitudinally underneath the aircraft. This produces a very narrow beam in a direction at right-angles to the flight path. The beam is wide in the vertical plane, usually cosecant^2 in shape, so that it covers a wide strip of ground. Each pulse from the radar scans the position of the s trip of ground covered by the antenna beam, echoes appearing as bright spots on a linear time-base on a high intensity CRO display. A movie camera synchronised to the aircraft speed photographs this display continuously thereby building up a radar map of the strip of ground. Figure 6.24 shows a diagram of the antenna beams; 2, one on each side, are shown but some systems have only one beam.

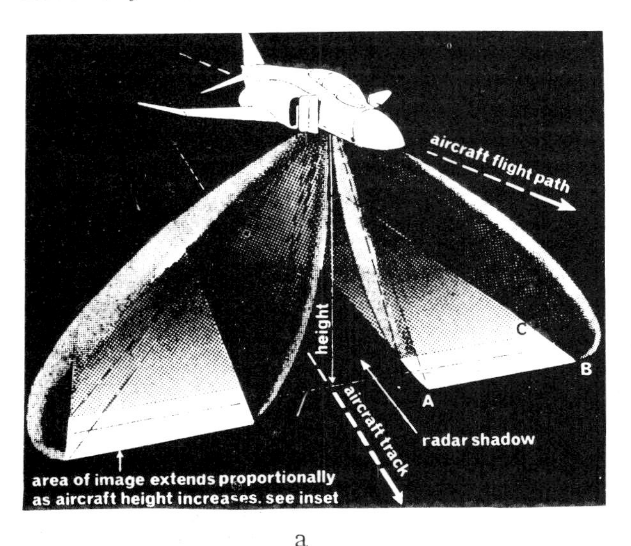

a

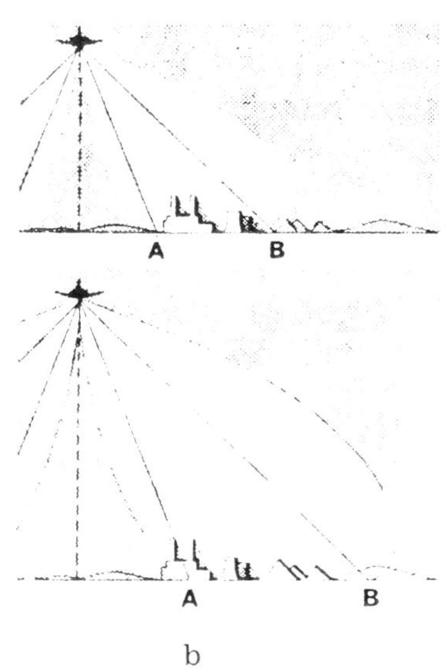

b

Fig. 6.24 SLAR

The radar usually has optional MTI so that it displays only targets moving on the ground (MTI system compensates automatically for the movement of the aircraft with respect to the ground). In order that the detail of the radar map shall be sufficiently fine, both the antenna horizontal beamwidth and the pulse width are very narrow; SLAR can resolve areas of ground 20 m x 20 m or better.

Airborne surveillance of airspace can extend the horizon to several hundred miles. Figure 6.25 shows a typical system, NIMROD in AEW configuration; the aircraft carries two radars, one in the nose and the tail, together providing all round surveillance.

Fig. 6.25 Nimrod in AEW configuration

TRACKING RADAR

A surveillance radar gives an approximate location of all targets within its surveillance volume or area. For some purposes this is enough, but a weapon system to destroy a target requires a continuous and accurate record of the location of this one target. Short range weapon systems often rely on optical target tracking, but longer range systems and systems with night time and all-weather capability require radar.

A tracking radar for airborne targets must provide continuous and accurate measurement of direction in both azimuth and elevation, and of range. Consider an antenna receiving with a narrow beam (Fig. 6.26). Because of the flat top of the beam position it is difficult for the radar to tell whether a target is at A, on the antenna axis, or at B or C, each some way off the axis and on opposite sides, and it could not track satisfactorily in angle. Figure 6.26 shows the beam split into 2 beams offset from the antenna, or boresight, axis by equal and opposite 'squint' angles. The radar angle tracking circuit takes the difference in target signal strength in the two beams; A, on axis, gives the same signal strength in each beam, hence the difference is zero; if the target is off axis at B, the signal in beam L is much greater than that in beam R, giving a large difference signal of one sign; a target off axis at C gives a large difference signal of the opposite sign. Figure 6.26 (c) shows this angle error characteristic. The angle error voltage drives a proportional control antenna position servo to keep the boresight axis on the target. By this means a good radar can track a target to within a twentieth of a beamwidth or better. Figure 6.26 shows tracking in 1 plane only; to track in the other plane as well requires a second pair of beams and a second angle error channel. This system of 2 pairs of coexistant beams is known as static split or 'monopulse'; monopulse because each received pulse gives a complete set of angle error information.

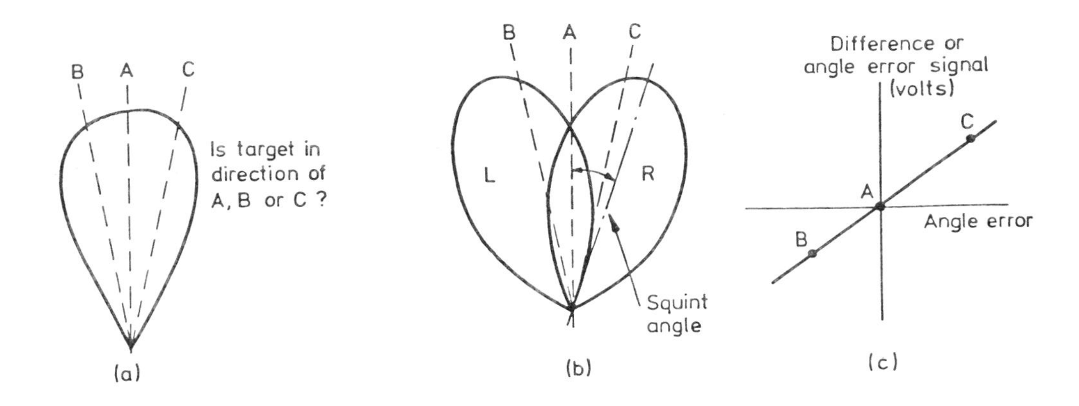

Fig. 6.26 Radar tracking in angle (one plane)

Static split is expensive and complicated though the quality of tracking is high. A simpler, cheaper system is to use a single offset beam which rotates about the boresight axis at a fairly high rate (up to 6000 rpm) (Fig. 6.27).

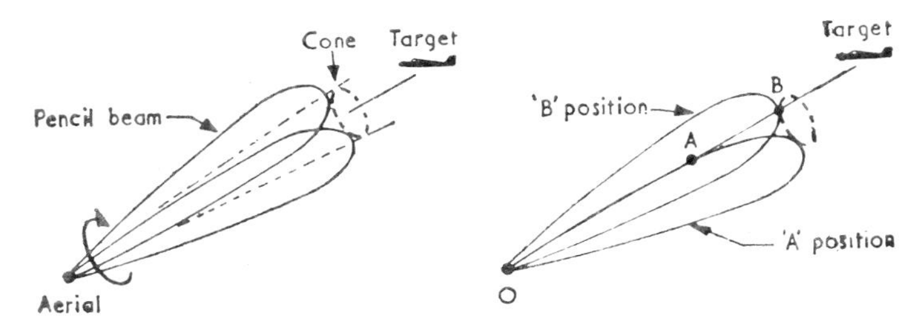

Fig. 6.27 Conical span

This single beam makes sequential measurements of target signal strength and thus obtains the same pair of difference or angle error signals. The system has the advantages that it requires a simpler antenna feed and a single channel receiver; on the other hand it is subject to errors arising from changes in target echoing area occurring during a period of rotation; it is also open to a deliberate form of interference known as spin-frequency jamming. A digital encoder on each of the 2 antenna mounting axes gives the corresponding target direction, in azimuth and in elevation for passage to a weapon control centre etc.

The tracking radar also measures range automatically. A servo system keeps the electronic range marker on the target echo, following changes in range. A digital data system also passes the range as represented by an internal servo system voltage to the same centre.

The presence of 2 adjacent targets is a potential source of difficulty, leading perhaps to complete loss of lock. Tracking radars always have a high degree of resolution, both in angle and in range, to minimise this difficulty.

IDENTIFICATION FRIEND OR FOE (IFF) & SECONDARY RADAR

One of the problems in radar is identifying a hostile target. It is sometimes possible to infer this from a target's behaviour, route, altitude, and from reports from other organisations. But an immediate and automatic way of establishing this is highly desirable. Radar-based IFF systems have been used since the Second World War, using a technique known as secondary radar. The Interrogator, which is linked to its parent radar, sends out a coded pulse train at a carrier frequency f_1, in the direction of the target which the radar has located but not identified.

Fig. 6.28 Secondary radar (IFF)

A friendly target carries a transponder which receives and decodes the interrogation; it then sends a coded reply on frequency f_2. The responsor receives the reply and, if it is correct, this indicates the target as friendly. Lack of a correct reply may be taken as meaning that the target is hostile, or, at least, unidentified. The current IFF Mk 10 operates in the 1 GHz frequency band with a 60 MHz separation between the 2 frequencies. The Interrogator sends out a train of 2 pulses whose interval is variable; the transponder replies in a code containing up to 14 pulses. The same system is used for civilian Air Traffic Control (ATC) though the system of codes is slightly different. The reply in this case contains information about identity, height, etc. and can be displayed on the radar PPI as shown in Fig. 6.29.

Difficulties in these systems arise from the confusion of replies from more than 1 transponder (garbling), and the over-interrogation of a transponder by several interrogators at once, leading to no reply at all. However circuits to mitigate these nuisances are usually incorporated. Another difficulty affecting IFF is that it is not 'fail-safe'. If the transponder of a friendly aircraft is out of action or is not working correctly, the defences may take the aircraft as being hostile. There is, as yet, no solution to this difficulty.

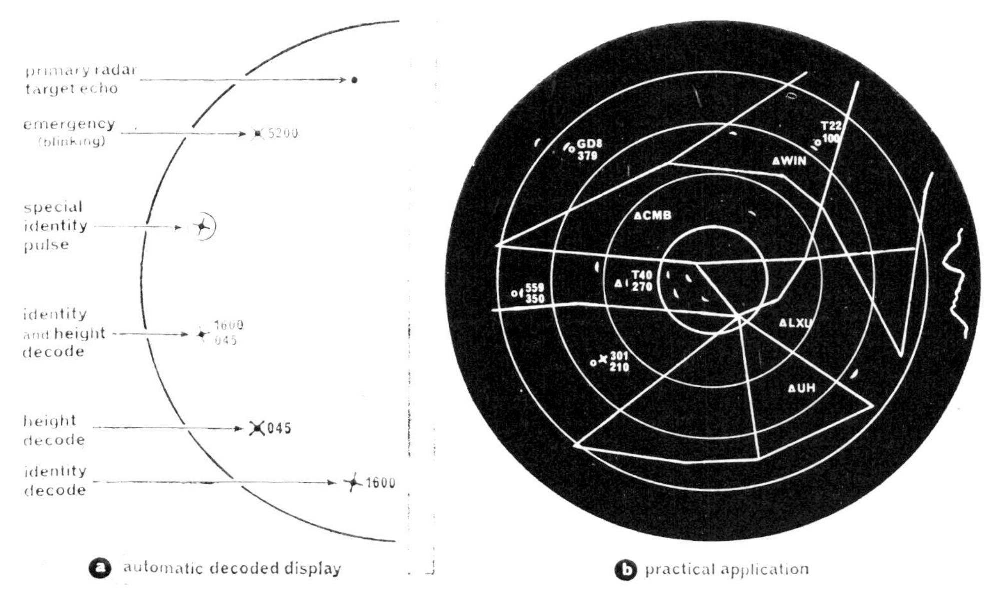

(a) automatic decoded display (b) practical application

Fig. 6.29 ATC display

Other applications of secondary radar are in navigation systems; for example an aircraft can locate itself by interrogating a network of fixed beacons and measuring the elapsed time of each reply. The idea can be extended to vehicles or even to men on foot. In surveying secondary radar has eliminated the tedium of measuring long distances by chain, giving an accuracy to a few centimetres.

ELECTRONIC OR INERTIALESS ANTENNA BEAM STEERING

There are obvious mechanical difficulties in turning a large antenna at a high rate, particularly if it scans a limited arc. It is also difficult to vary a scanning programme; the system is inflexible. Electronic beam steering eliminates the need to move any part of the antenna mechanically; it thus offers much faster scan rates, scanning programmes can be varied at will and it is even possible to combine the tracking of individual targets with continued surveillance. The major limitation is that it is not possible to scan a sector of more than 120° so that all-round coverage requires three, or preferably four arrays.

Figure 6.30 shows the principle of electronic beam steering in 1 plane. A conventional single plane array (a) consists of a line of individual radiating or receiving elements, all of which act in phase. The array transmits or receives a wavefront which is parallel to the line of the array so that the direction of propagation or reception is broadside. Introducing a progressive change of phase

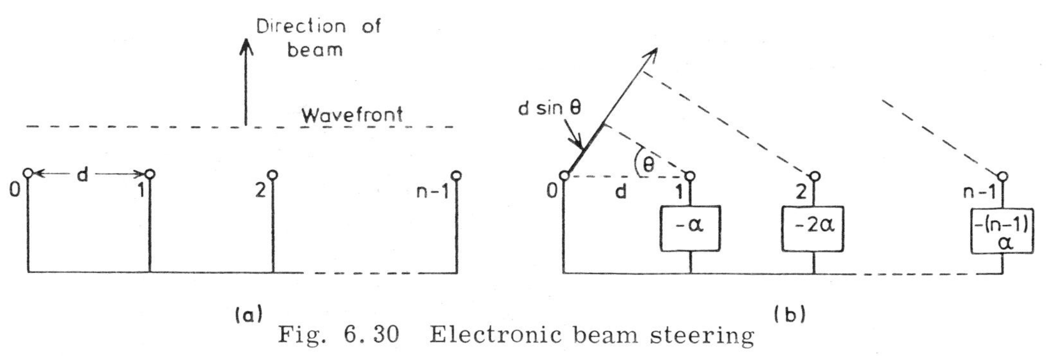

Fig. 6.30 Electronic beam steering

along the array (b) causes the beam to turn. Say that the progressive change of phase is - α radians from element to element, in the wavefront common to elements 0 and 1 the radiation must be in phase along the wavefront; to compensate for the phase retardation of α from 1 the radiation from 0 must travel an extra distance d sin θ so that the correspodning phase retardation, $2\pi\ d \sin\theta/\lambda$, is equal to α. Hence the wavefront and the direction of the antenna beam are turned through an angle $\sin^{-1}(\alpha\ \lambda/2\ \pi d)$. Reversing the sign of α turns the beam in the opposite direction.

This is known as a 'phased array'; the phase shifters are electronic devices which act very quickly giving rapid movement of the beam. By controlling the phase shifters from a computer elaborate scanning and tracking programmes are possible. The diagram shows steering in a single plane, steering in both horizontal and vertical planes is possible by using a planar array of several rows or columns of linear arrays. Ideally each element in the array is controlled separately but with, say, 10,000 elements; this is no mean task. A simpler but less versatile system controls the phase of each row and of each column of radiators. Figure 6.31 shows the planar phased array of a PATRIOT 3D air defence radar which combines surveillance with accurate tracking of several targets simultaneously.

Fig. 6.31 A phased array

An extension of electronic beam steering is to generate a number of beams in fixed directions by a matrix of fixed phase shifters and coupling devices. This technique is used for elevation measurement in air defence surveillance radars with conventional mechanical scan in azimuth, and for covering a sector in azimuth with a battlefield surveillance radar. Multiple beam formation can also be combined with electronic beam steering to generate a moveable cluster of 4 beams for tracking a target by static split.

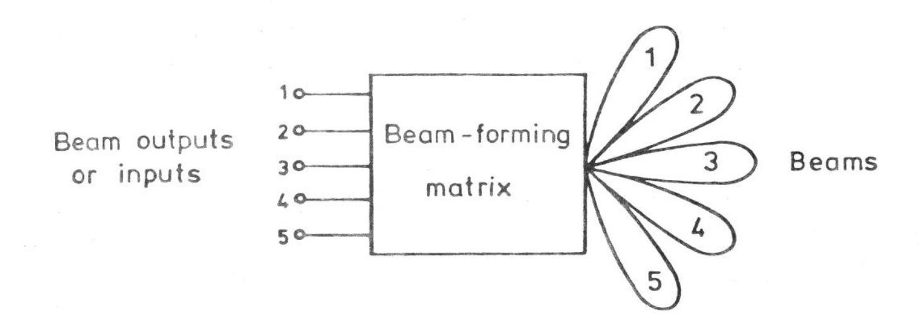

Fig. 6.32 Multiple beam formation

Phased arrays are expensive and the need for 4 arrays for all-round surveillance is a serious disadvantage. A compromise is to combine conventional mechanical rotation in azimuth with electronic beam steering in elevation, Fig. 6.33 shows such an array.

Fig. 6.33 Electronic beam steering in elevation

Phased arrays will probably never be as cheap as their mechanical counterparts but the development of cheap lightweight printed-circuit antennas, the large-scale production of miniature phase shifters and the advent of the microprocessor promise a more widespread use of electronic beam sterring.

PULSE COMPRESSION

This is another important modern development. Good range resolution calls for a correspondingly short pulse; however detection range depends upon the average transmitted power, hence the shorter the pulse the greater the pulse power to maintain the required average power. In large radars with valve transmitters the pulse power available, though high, is limited by breakdown of insulation by the high voltage required and inability of the valve to generate the high current; it is also limited by electrical breakdown in the feeders and by corona at the antenna. In small radars with semi-conductors it is limited by breakdown in the material. There is, therefore, a potential incompatibility in the two requirements.

The solution to the difficulty is to radiate a much longer pulse at a correspondingly lower pulse power and to restore the range resolution by modulating the radio frequency carrier within the pulse envelope. The modulation must have a bandwidth B equal to the reciprocal of the equivalent time resolution. Say that our radar with a resolution of 15 m had to radiate a pulse of not less than 5 μs to achieve the necessary average power. To restore the range resolution the modulation of the carrier must have a bandwidth of 10 MHz; a common form of this intra-pulse modulation is linear FMCW (Chirp) and in this case the carrier frequency would increase, or decrease, linearly from the beginning to the end of the pulse (Fig. 6.34).

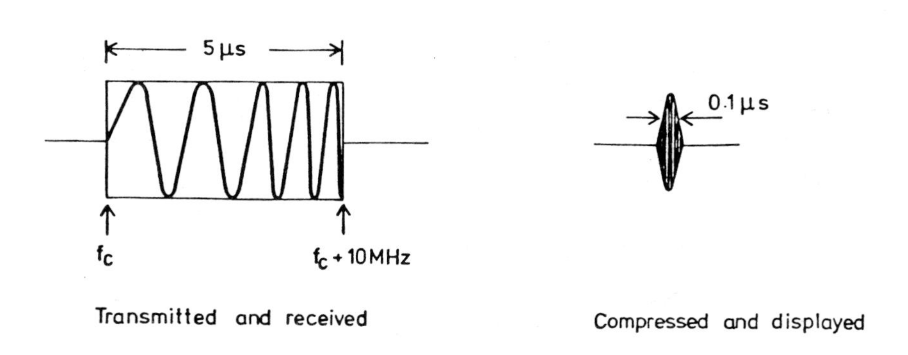

Fig. 6.34 Pulse compression

Each received pulse is of the same form as this but the receiver contains a 'pulse compression' network which is matched to, or 'recognises' the modulation and compresses the pulse to the required width of 0.1 μs. Other types of modulation produce the same effect provided that they have the necessary bandwidth but the only other type in general use is coded phase modulation. Difficulty may arise in pulse compression if there is a Doppler frequency shift on the echo; however, provided that the bandwidth of the modulation is much greater than the Doppler shift the effects are not serious.

USE OF MILLIMETRE WAVES

These very short wavelength waves are attractive because they need much smaller antennas to generate a narrow beam and hence to give good angular resolution. For example, a cylindrical beam of width 10 mils at a frequency of 94 GHz requires a paraboloid of diamter only 15", whereas at a 'normal' radar frequency of 9.4 GHz the paraboloid would have to be 12½' in diameter!

The principle objections to the use of millimetre waves are the high atmospheric and weather attenuation, and the lack of convenient high power sources. Millimetre wave radars utilise 'windows' in the atmospheric attenuation graph, eg at 35 GHz (8 mm), 94 GHz (3.2 mm) and possibly 140 GHz (2.15 mm). Nevertheless the attenuation is still high, a factor of 6 over a range of 10 km at 94 GHz, though not so high that operation in clear weather is impossible. Over the same distance in moderate rain the factor is about 100,000, showing clearly the limitations of millimetre waves.

High power valve sources are available but they are inefficient and require bulky and inconvenient ancillaries such as high voltage power supplies and powerful magnets. These nullify any advantage accruing from the small size of antenna. Solid state sources are compatible with the concept of a miniature radar but, at present, powers available are only a few hundred milliwatts, sufficient for short ranges only.

Nevertheless there are practical applications for mm wave radar in short range high definition surface target surveillance, for intruder alarms and for close support air defence radars. The high resolution possible with mm waves makes them particularly suitable for ground mapping radars; an airport ground movement surveillance radar even depicts the shape of aircraft, though it cannot, as yet, identify the airline! mm waves also offer comparative freedom from electronic countermeasures; jammers have to cover a much wider band of frequencies than at longer wavelengths, furthermore they are hindered by a lack of suitable sources. However, as in all matters pertaining to electronic warfare, it would be wrong to be complacent.

A rather different potential application of mm waves is in 'passive radar'. This relies upon comparison of the thermal emissivity of a target with that of its background; it is the radio equivalent of thermal imaging. Unfortunately the thermal radiation at such long wavelengths is very much less than in the IR band (by a factor of about 100,000) and the difference is very difficult to detect. Nevertheless aircraft have been detected at a few kilometres against the background of a 'cold' sky and tanks at a few hundred metres against the much 'hotter' background of vegetation. This is a promising technology but advances depend upon the improvement of the sensitivity of the passive receiver, or radiometer.

SELF TEST QUESTIONS

QUESTION 1 Radars usually operate at much shorter wavelengths than conventional radio because:

(a) It is easier to generate the large transmitter power needed;
(b) Range accuracy is better;
(c) Narrow antenna beams are more practicable;
(d) Penetration of the atmosphere is better.

Answer ....

QUESTION 2 a. Calculate how long it takes an echo to return from a target at a range of 25, 550 metres.

b. Estimate the minimum pulse duration for the radar to resolve a second target 50 metres further away.

Answer a. ....

....

b. ....

....

....

QUESTION 3 The p.r.f. of a pulse radar with an operational range of 50 km should be:

(a) Equal to or less than 6 KHz;
(b) Equal to or greater than 6 KHz;
(c) Equal to or less than 3 KHz;
(d) Equal to or greater than 3 KHz.

Answer ....

QUESTION 4 Match the antenna beam shapes (A-D) with the rôles (W-Z).

A	beaver-tail	W	plan-position indicating
B	cosecant^2	X	weapon locating
C	narrow fan	Y	elevation coverage
D	pencil	Z	height-finding.

Answer ....

QUESTION 5 a. Why can a falt metal plate (area $>> \lambda^2$) give a very large echo?

b. What sort of echo does a raindrop give?

c. How does the echo from a thin metal rod depend upon the plane of polarisation?

d. Why does a metal cone viewed nose-on give a very small echo?

(a. and d. can be checked with light).

Answer a. ....

....

....

b. ....

....

....

c. ....

....

....

d. ....

....

QUESTION 6 Which of the following statements pertaining to surveillance radar are true?

(a) Doubling the antenna linear dimensions increases antenna gain 4-fold.
(b) Doubling the antenna linear dimensions doubles the beamwidths.
(c) Doubling the antenna linear dimensions doubles the detection range.
(d) Halving the wavelength has the same effects as doubling the antenna dimensions (ignore atmospheric and other effects not related directly to the antenna).

Answer ....

QUESTION 7 The radar echoing area of a target T2, which is at a range of 50 km and which gives an echo of the same strength as a target T1 at 20 km, is:

(a) 2.5 times greater than that of T1;
(b) 6.25 times greater than that of T1;
(c) 39 times greater than that of T1;
(d) 1526 times greater than that of T1.

Answer ....

....

....

QUESTION 8 Calculate the speed of an approaching vehicle in mph as observed by a police radar which records a Doppler frequency of 1.35 kHz. The radar wavelength is 30 mm, 1 $msec^{-1}$ = 2.23 mph. Is it certain that this is the true speed; if not, give reasons.

Answer ....

....

If the radar were misaligned accidentally so that the radar signal was reflected off a wall parallel to the road would this

(a) favour the motorist;
(b) favour the police;
(c) make no difference?

Answer ....

QUESTION 9 a. Which of the following is most probably true? Tracking radar can track to an angular accuracy of:-

(i) 1 to 2 beamwidths;
(ii) $\frac{1}{2}$ a beamwidth;
(iii) 1/20th to 1/10th of a beamwidth;
(iv) 1/100th of a beamwidth or better.

Answer ....

b. Suggest a value for the maximum permissible tracking error for a radar which is guiding a missile to hit a target 20 metres square at a range of 6 km.

Answer ....

....

QUESTION 10 State the connotation of each of the following:-

(a) blind speeds
(b) pulse compression
(c) phased array.

Answer (a) ....

(b) ....

(c) ....

ANSWERS ON PAGE 194

7.

Surveillance in Depth

SURVEILLANCE IN DEPTH

> "I have spent all my life in trying to guess what lay on the other side of the hill". The Duke of Wellington.

For centuries the battlefield commander has been concerned about events beyond the limit of his visibility. Even in the era of the musket and the sword he needed to know about the approach of enemy forces, their strengths and their dispositions. Without such vital information his ability to react effectively to the threat was seriously reduced.

The modern battlefield commander has at his disposal weapon systems with far greater range and firepower than those of his predecessor. The area of influence of modern armies over the battlefield has been extended well beyond the line of sight by such weapons. The commander will need the means to detect and identify the build-up of enemy forces at an even greater range in order that he can pre-empt, if possible, the use of such long range weapons on his own forces. It is about systems that afford the commander this surveillance and target acquisition capability that this chapter is concerned.

High resolution surveillance systems using advanced electronic and electro-optic techniques have given modern armies a near 24 hour all-weather surveillance capability. These developments have accentuated the limitations on surveillance in depth imposed by the curvature of the earth, terrain, natural vegetation and man-made obstacles. In North West Europe, for example, the average limit of visibility (not taking weather conditions into account) is about 4 or 5 kilometres. Therefore, the only means of extending the range of surveillance systems is to raise sensors above the ground or deploy them well forward into enemy territory (Fig. 7.1).

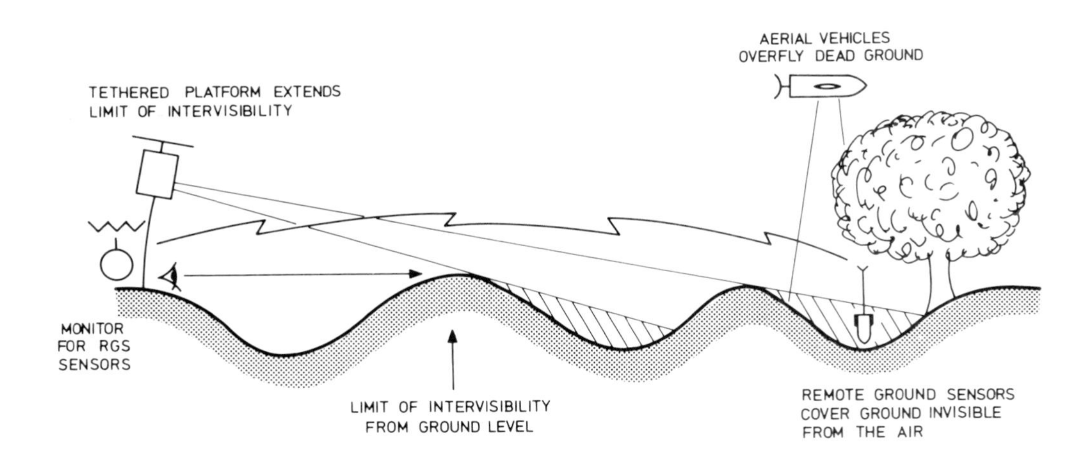

Fig. 7.1 Surveillance sensors

The use of high ground has long been recognised as a means of dominating the enemy. From the earliest times fortifications have been built on high ground covering main routes and approaches. Although the potential of aerial surveillance must have been appreciated centuries before, it was not until the French deployed a hot air balloon at the Battle of Fleurus in 1794 that aerial observation was first used. Tethered observation balloons were widely used in the First World War but they soon became too vulnerable to enemy action and the development of the aircraft heralded the birth of air reconnaissance as we know it today. The manned aircraft remained the principal platform for airborne surveillance sensors until well after the Second World War. However, the increase in effectiveness of air defence systems due to the advent of fire control and missile systems and the increasingly high cost of sophisticated manned aircraft has led to the diversification of surveillance techniques. Furthermore, the penetration of space in the 1950s made the surveillance of the earth's surface from orbital satellites a possibility.

SURVEILLANCE FROM SPACE

Since man first inserted a satellite into orbit around the earth military and scientific observation of the earth's surface and atmosphere have become commonplace. The development of photographic lenses capable of detecting objects less than a metre in size at altitudes typical of satellites in orbit around the earth has given nations capable of launching them an unequalled surveillance capability over much of the earth's surface. The use of infra-red sensors in particular allows the detection of activities of military significance such as the launching of missiles and the detonation of explosive materials. One of the great advantages of surveillance from space has been the shift of emphasis away from strategic reconnaissance by U2 type aircraft that have in the past been the subject of so much international political controversy. Indeed, the monitoring of military activities from orbital satellites has become a recognised means of arms control between the superpowers. However, the use of space for tactical surveillance of the battlefield is not yet as cost-effective as other systems available. The cost of inserting satellites into orbit is prohibitive and the low orbital trajectories required for optimum surveillance conditions implies a relatively short satellite life. Furthermore, the ground coverage of surveillance satellites is limited and a large part of their orbit is of no use in the tactical application. Even when using infra-red sensors, satellites are largely ineffective when weather conditions are very poor. Nevertheless, satellites have been launched by the superpowers for tactical surveillance for monitoring purposes during limited wars in the Middle East. Their role for surveillance in depth must therefore be recognised but so too must their limitations, and in particular cost.

ELEVATED SENSORS

The remainder of this chapter is devoted to the examination of surveillance systems deployed on the ground or within the atmosphere.

Several methods of raising sensors above the ground to get a better angle of sight have been tried since the days of balloon observation. For example, kites have been tried but their inherent instability, their reliance on the vagaries of wind speed and direction and their limited payload rule them out as a satisfactory sensor platform.

The Swedish GIRAFFE, an air defence surveillance radar with a range of 20-40 kilometres, has an antenna mounted on a 12 metre high folding mast which not only increases its range but also allows deployment in wooded areas.

Tethered Platforms

Dornier, of West Germany, has produced the KIEBITZ (Fig. 7.2) which is a rotor powered tethered platform on which can be mounted a variety of sensors, including the Franco-German ARGUS surveillance radar, Low Light Television (LLTV), IR and EW Devices.

The manufacturers claim to have overcome the problem of stabilisation of sensors and the platform can be flown in steady winds up to about 15 metres per second. KIEBITZ has an impressive endurance since fuel is supplied to the gas turbine engine through the tethering cable which is also used for the transmission of control, monitoring and sensor signals. The 8 metre diameter rotor is driven by the ejection of compressed air at the blade tips and this reaction propulsion system eliminates the need for a tail rotor. The platform is flown at an altitude of 300 metres which gives the radar a range of up to about 50 kilometres depending upon the terrain. The same company has also developed an ingenious unpowered tethered rotor platform called the SPÄHPLATTFORM of 1.2 m diameter, which is vehicle mounted and uses the vehicle's engine to accelerate the rotor to a high speed on the ground. The ring surrounding the rotor acts as an energy store and gives the device an endurance of just over one minute during which it carries optical or electro-optical sensors to an altitude of 100 metres. This device is well suited for use in armoured vehicles deployed close to the FEBA for limited surveillance in depth.

Fig. 7.2 KIEBITZ

Long Range Airborne Systems

It is appropriate to mention two aircraft mounted surveillance systems designed to operate on the near side of the FEBA. The first is the now well known Airborne Warning and Control System (AWACS) deployed by both NATO and Warsaw Pact forces for the detection of airborne targets as an early warning for air defence systems. Typically AWACS consist of high powered pulse doppler radars fitted to aircraft flying at about 10,000 metres and able to detect airborne targets down to ground level at ranges in the order of 500 kilometres. This type of radar also has a limited capacity to detect mobile ground targets and surface ships. The principle was to have been employed in the American Stand Off Target Acquisition System, SOTAS, which until its cancellation in 1981 was under development for use at divisional level for depth surveillance and the acquisition of targets (Fig. 7.3).

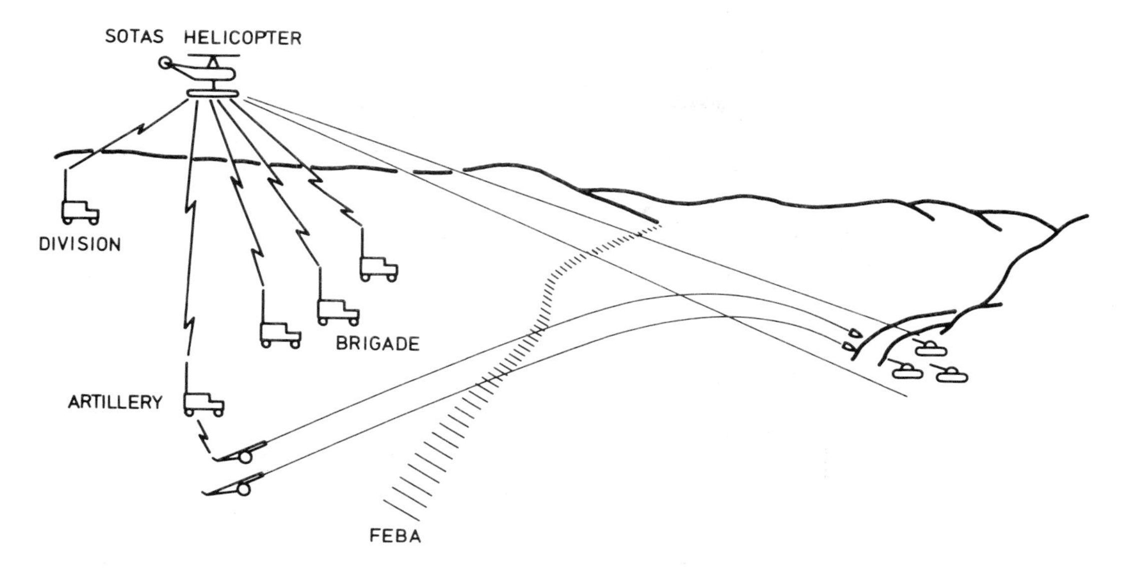

Fig. 7.3 The SOTAS concept

A typical stand off surveillance and target acquisition system consists of a helicopters carrying long range radars flying well out of range of enemy air defence systems on the near side of the FEBA. The radars scan either a sector or through 360° and detect moving targets well beyond the FEBA at ranges in the order of 50-80 kilometres from the aircraft. Target imagery is transmitted via a data link to ground stations located at divisional, artillery and brigade HQs where information is displayed on map related displays in near real-time. This type of system is ideal for general surveillance of the divisional Area of Influence to allow the acquisition of targets for long range artillery and airborne weapons. It may also be used in conjunction with other surveillance and target acquisition systems with better resolution for recognition and identification of the nature of the target.

So far in this chapter we have looked at long range surveillance and target acquisition systems that are ground-based, elevated or mounted in aerial vehicles deployed on the near side of the FEBA. This renders them relatively invulnerable to enemy action but their resolution is often not fine enough at longer ranges to provide recognition and identification of targets necessary for discriminatory engagement by long range weapons. Furthermore, even when sensors are mounted in aircraft, targets can be screened by terrain and other obstacles at longer ranges. To overcome these problems the only solution is to deploy sensors on the ground or in aerial vehicles forward of the FEBA.

UNMANNED AIRCRAFT

Survivability has become one of the primary considerations in the design of surveillance systems deployed forward of the FEBA. The proliferation of air defence systems at all levels and the very high cost of manned aircraft has rendered them too vulnerable for use as the primary means of forward airborne surveillance.

Unmanned aircraft provide the best answer to the requirement for timely combat intelligence in depth. Under the direct control of the army commander they can be tasked quickly to respond effectively in a short time from positions within the divisional area. The size of unmanned aircraft can now be reduced to make them difficult to detect and destroy. Miniaturisation of powerplants and avionics and the advent of microprocessors has helped in this reduction of size.

Unmanned aircraft are either classified as drones, which fly on pre-programmed flight paths requiring no communication link with the ground (although some systems do employ a real-time sensor data link), or remotely piloted vehicles (RPVs) which are controlled either from the ground or from a manned aircraft over a command data link.

DRONES

The drone is generally less sophisticated than the RPV and is more widely deployed in modern armies. A typical drone is the Canadair CL 89 currently in service with the British Army where it it known as the AN/USD 501 MIDGE (Fig. 7.4).

MIDGE is deployed in support of divisional formations and provides surveillance in depth out to between 40 and 60 kilometres depending on its task. Requests for drone reconnaissance missions over specific areas are initiated by the divisional intelligence staffs and orders are passed to the drone troop for immediate planning of the sortie. MIDGE carries either a photographic or infra-red sensor which can be used in completing a variety of missions. The quickest reaction time is achieved for confirmatory tasks when the sensor is switched on for sufficient time to obtain cover of a particular point, eg a bridge to confirm that it is intact. It can be programmed to carry out a strip search on a straight flight path between 2 points; the longer the strip the longer the reaction time due to the amount of interpretation of film that is required. It can also be tasked to carry out reconnaissance of an area by a systematic search strip by strip, but MIDGE is not ideally suited to this type of mission due to the mass of photographic material produced and limited facilities within the photographic processing and interpretation vehicle.

At the end of the mission the drone is guided to the recovery area by a homing beacon and it is recovered by parachute. Cushioning bags are deployed under the drone to prevent damage on hitting the ground. It is vital that film is recovered as quickly as possible and processed in the photographic processing and interpretation vehicle which is sited adjacent to the recovery area. The results of the mission are then sent by the quickest means, usually a radio link, to the divisional intelligence staff who initiated the request.

A typical response time, which is defined as the time from the initiation of a request for information to the time at which the information is received by the initiator, is about one hour. It is this fast response time, an ability to operate from

Fig. 7.4 The MIDGE drone at launch

small unprepared sites and invulnerability to electronic warfare that are the main advantages of the drone. Furthermore, the divisional commander, to whom surveillance in depth is so vital, has direct control over them. At the same time it is important to recognise their limitations which include a limited radius of action, a pre-programmed flight path which cannot be altered after launch and the intelligence gathered is nearly one hour out of date when it is sent to the originator. Other drone systems, such as the Canadair CL 289 (AN/USD 502), do incorporate a real-time data link to overcome this disadvantage, but this is bound to increase the cost and complexity of the drone and renders it vulnerable to electronic countermeasures.

A majority of drone systems are launched from a ramp in unprepared launching sites in the combat zone. Some are launched using a conventional landing gear from prepared sites and one or two are released from aircraft. Several methods of recovery are in use although parachute or net are the most common means of retrieval.

REMOTELY PILOTED VEHICLES

RPVs can overcome some of the inherent limitations of drones. By definition they are remotely controlled by a ground based or airborne 'pilot' and are consequently able to react to commands during flight, usually as a result of intelligence gathered by their own sensors from which information is transmitted over a real-time data link. Although this link is not strictly essential the major advantages of the RPV over the drone would be lost without the information 'loop' created by the remote control and data links.

Roles

Because of the greater flexibility of the RPV it can be used for more diverse roles than the drone. It is ideal for general surveillance using high resolution sensors such as television or Thermal Imagers (TI). Any potential targets detected by the controller on his visual display can be investigated either by altering the field of view of the sensor or by manoeuvering the RPV into a better position to investigate. Once the RPV has detected a target and the controller has identified it as suitable for engagement by artillery fire the RPV can loiter overhead to adjust the fire onto the target. For this purpose it is necessary for the sensor to have a large enough field of view to detect the first round of artillery fire which, because of the zone of indirect fire weapons at long ranges, could fall within an area as large as 400 or 500 metres square. Once the initial round is seen the controller should be able to give orders to adjust the fire onto the target on the second round. By manoeuvering the RPV directly over the target the controller can obtain a plan view of the ground from which corrections can be measured with great accuracy. With the advent of the Cannon Launched Guided Projectile (CLGP) it is feasible for the RPV to carry a laser designator with which to illuminate the target to assure a first salvo hit by CLGPs. It is necessary, of course, for RPVs operating in this role to be fitted with a high resolution imaging system in order to identify and locate targets for laser illumination. Lockheed Missiles and Space Co. is developing a mini RPV for the US Army in just this role. The RPV has an inertial navigation system which can be programmed, before or during flight, with the coordinates of way points to which the aircraft is to fly (which might, for example, be obtained from other surveillance systems such as SOTAS). Once the RPV is over a point of interest the operator can command the sensor into an "auto track" mode to hold it on target regardless of the motion of the RPV. The laser can then be fired in the range finding mode and the coordinates of any target within the field of view of the TV sensor are indicated automatically in the ground control station. The laser can also be operated in the designation role for use with terminally guided munitions such as CLGP. Typical battlefield tasks for RPVs are shown in Fig. 7.5.

RPVs may also be used for Electronic Warfare (EW) as a platform for Electronic Counter Measure (ECM) instruments. In general, the closer these devices are deployed to the enemy the more effective they are. RPVs, with their higher level of survivability than manned aircraft, are ideally suited and cost effective in this role.

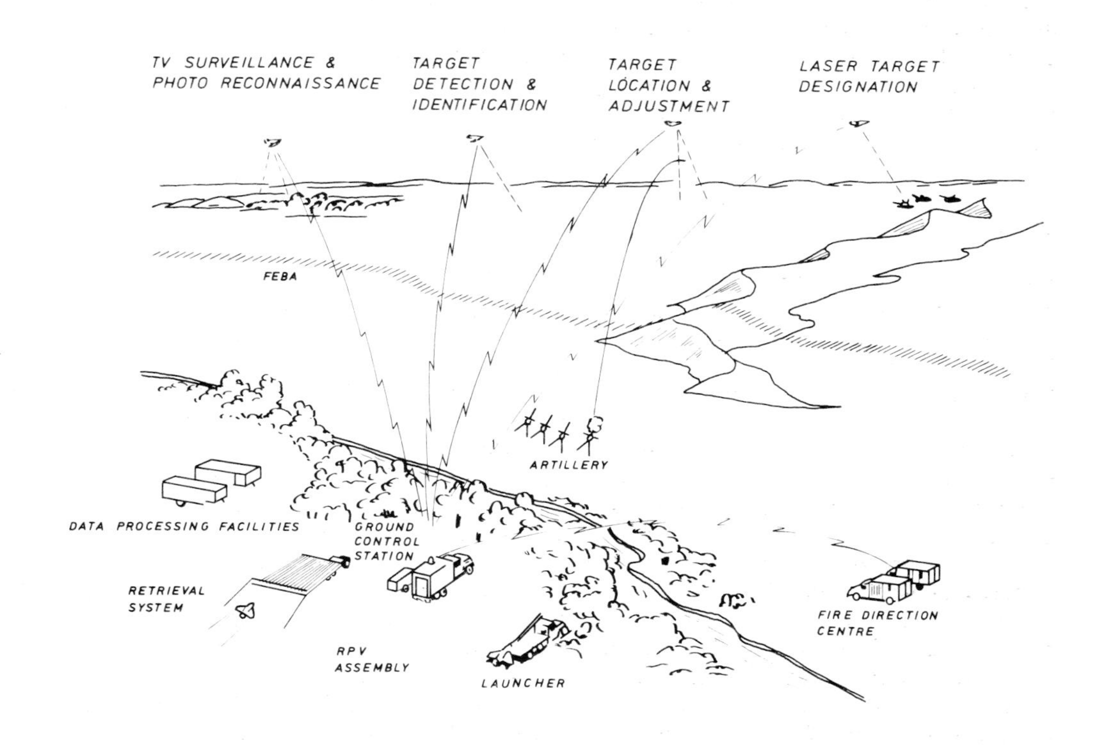

Fig. 7.5 Battlefield tasks3

Remotely Piloted Helicopters (RPH)

Although the majority of RPV systems so far developed use fixed wing aircraft a few mini helicopter, or RPH, systems have been produced. The majority of these air vehicles have plan symmetric shapes with contra-rotating rotors which allows them to fly in any direction without the need to alter the orientation of their fuselages. It also allows dtaa link antennas to remain directed at the ground control station and the on-board sensors can be kept in the same orientation: for example North can be maintained at the top of sensor displays. This is difficult to achieve in fixed wing RPVs. RPH systems can also be launched and recovered in virtually any type of terrain which is an advantage in forward areas of combat zone. However, the helicopter is less efficient than the fixed wing aircraft and

Fig. 7.6 The Canadair CL 227 RPH

their speed and endurance is not as good. The Canadair CL 227 is an example of an RPH (Fig. 7.6) which resembles an upright dumb-bell with twin three-bladed rotors on the inner section. The upper sphere houses the engine and fuel supply and the lower sphere contains the sensors.

Data Links

One of the major operational difficulties of the RPV is the maintenance of an unobstructed line of sight between its data link antenna and the control station. The higher the frequency of this link the more critical is the requirement for an unobstructed path between them. The choice of frequency is governed by the amount of information that the link is required to carry. TV and Thermal Imaging (TI) sensors require a broadband or high information rate link and therefore higher frequency than Infra-red Linescan (IRLS) or Moving Target Indication (MTI) radar sensors. Clearly the lowest possible frequency consistent with the required data rate should be chosen to reduce this problem. Accidental loss of the link does not necessarily lead to the loss of the RPV. Its navigation system can be programmed to initiate emergency action in this eventuality, eg by returning to the point at which communication was lost or by increasing altitude. A deliberate loss of link might even be operationally desirable on a pre-programmed mission although the loss of real-time sensor information is an obvious disadvantage. It is important that the role of the RPV is clearly defined and only the sensors required to meet its role are carried. The principal elements of an RPH and control station are shown at Fig. 7.7.

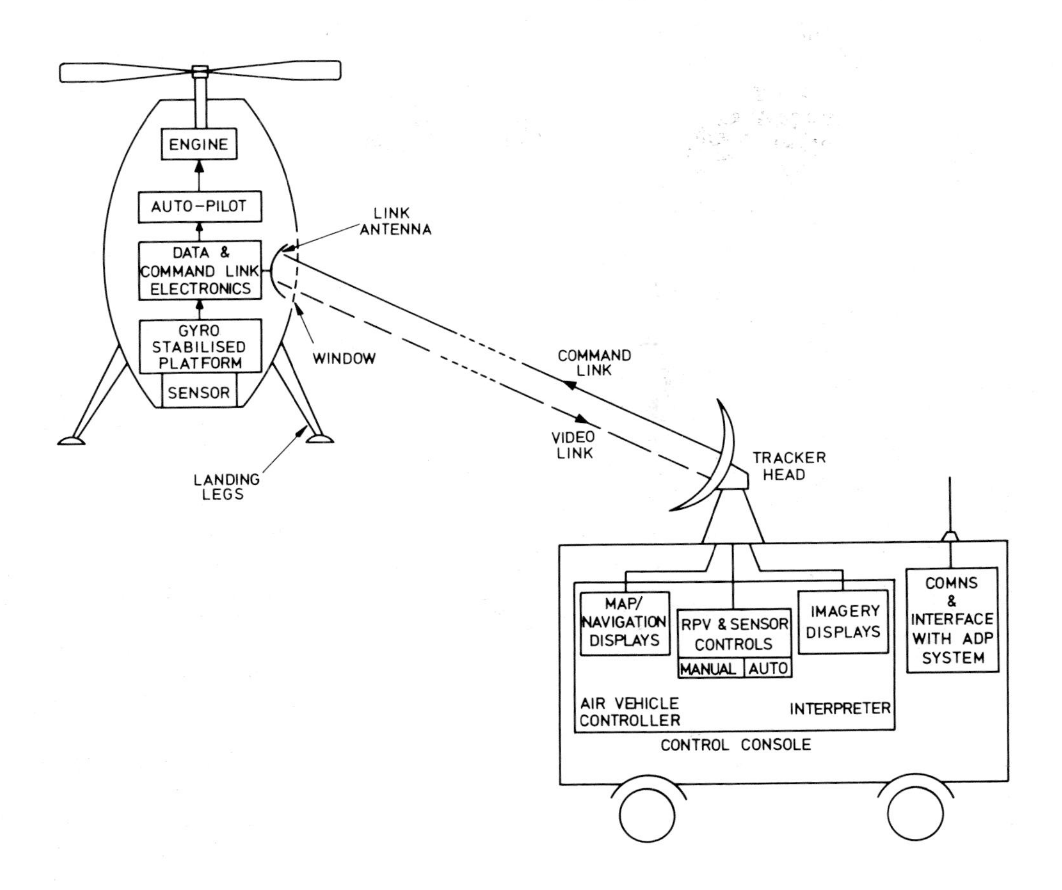

Fig. 7.7 Principal elements of RPH and control station

Costs

It is important that the cost of an RPV system is kept in perspective vis-a-vis the cost of manned aircraft for their justification must be largely based upon this financial ratio. By the very nature of the roles of RPVs, and in spite of their relative invulnerability, loss rates will be higher than can be tolerated for manned aircraft. It follows therefore that if the cost of an RPV is higher than a certain proportion of the equivalent manned aircraft its credibility must be in doubt. The cost of producing the basic structure of the vehicle is roughly proportional to weight; therefore airframe costs should be acceptable. It is the cost of engines, avionics and sensors which must be kept down and it is important that only those required to meet the defined operational requirements of the RPV are incorporated. Conversely, it is vital that the user of the system is aware of the technical and financial implications in defining required capabilities.

REMOTE GROUND BASED SYSTEMS

The Need

So far we have looked at surveillance and target acquisition systems with some limitations on their ability to maintain 24 hour surveillance over the battlefield by virtue of terrain screening, limited endurance or bad weather conditions. In order to close this gap in his capability the commander must deploy troops close to and even beyond the FEBA. With the aid of image intensifiers, thermal imagers and battlefield radars, forward and deep penetration patrols can keep a 24 hour watch over areas out of reach of other surveillance devices, particularly those in dead ground or screened by natural vegetation. However, these patrols absorb manpower which is often not readily available and other means of covering gaps have to be found. Remote Ground Sensors (RGS) are ideal in this role.

Remote Ground Sensors

Intruder alarms based on seismic and infra-red sensors have been in service with many armies for some time. They have a useful role in the close protection of troops and installations by warning of the approach of men or vehicles. Usually seismic detectors buried in the ground or infra-red sensors are connected to monitors by short lengths of cable or wire. However, it was not until the late 1960s in Vietnam that the Americans first exploited the potential of remotely emplaced sensors relaying information to monitors over radio links. Supply routes along the Ho Chi Minh Trail were made up of a vast network of secret paths through dense vegetation which were virtually impossible to detect from the air. In order to monitor activity along these trails the Americans developed a range of hand emplaced or air deployed sensors which, when they detected the presence of men or vehicles, sent signals over radio links to monitors based either on the ground or in aircraft which patrolled overhead. Long range artillery or air strikes were then directed at the locations where movement had been detected. Since then modern microelectronic and signal processing techniques have been applied to RGS design and complex systems, such as the American Remotely Monitored Battlefield Sensor System (REMBASS), are being designed and developed.

Uses

The purpose of a RGS system is to detect, locate and possibly identify targets. Earlier intruder alarms could not carry out these functions satisfactorily mainly because they were unable to discriminate between wanted and unwanted signals. The requirement is for a simple, reliable device with a low false alarm rate, small enough for concealment and preferably with an anti-tamper or self-destruct facility. The sensor signal must be processed and transmitted over a data link, in some cases over extended ranges requiring relays, to a monitoring station. It should ideally only transmit when a significant event has occurred, and should reveal the number of men/vehicles forming the target, their speed and direction, and the vehicle type. The complete system requires a number of sensors at known positions with each identified by an electronic coding device. The basic technical problem is to identify a characteristic generated by a target which is detectable by the sensor at the required range and still retains the identifying features of the emitted signal.

Sensors

RGS systems employ a variety of sensors each designed to detect target emissions at different wavelengths in the electro magnetic spectrum. REMBASS, for example, employs 3 different types of seismic/acoustic sensors. One is hand emplaced, the second designed for delivery from an aircraft or helicopter and the third is fired from an artillery piece such as the 155 mm howitzer. They have a detection radius of 500 m for vehicles and 50 m for personnel and can classify targets into tracked or wheeled vehicles. The system includes hand emplaced magnetic and infra-red sensors both of which are capable of determining the direction of travel of the target by the deployment of 2 sensors which need to be placed about 5 metres apart.

Communications Links

In order to achieve surveillance in depth using RGS systems it is important that repeaters are deployed to relay the sensor signals to monitors situated in friendly territory. REMBASS includes repeaters that can be placed by hand or delivered from the air, the latter having ground implanted antennas or, in another version, antennas that are designed to hang up in trees for greater range.

Monitors

The third part of the RGS system is the monitor which receives signals direct from the sensor or via repeaters. The signals are processed and displayed, indicating which sensor has detected movement, the nature of the target and when and how many were recorded over what period. Data can be displayed electronically and at the same time printed out for a permanent record. Alarm systems can be built in to alert the operator when information is received. Clearly each RGS system can be designed to present information in different ways. Their effectiveness has certainly been improved by recent advances in signal processing and miniaturisation of components. In particular the use of microprocessors programmed to recognise target patterns has helped to increase effectiveness and reduce false alarms. However, the effectiveness of the system must also depend upon careful siting of sensors to help in detecting build up of enemy forces in depth, or their approach towards defended locations. Whilst limited recognition of targets is possible using RGS systems they are intended more as an alerting device to trigger the deployment of other surveillance and target acquisition systems with means of more specifically identifying the nature of the threat. Figure 7.8 shows a schematic layout of a RGS system.

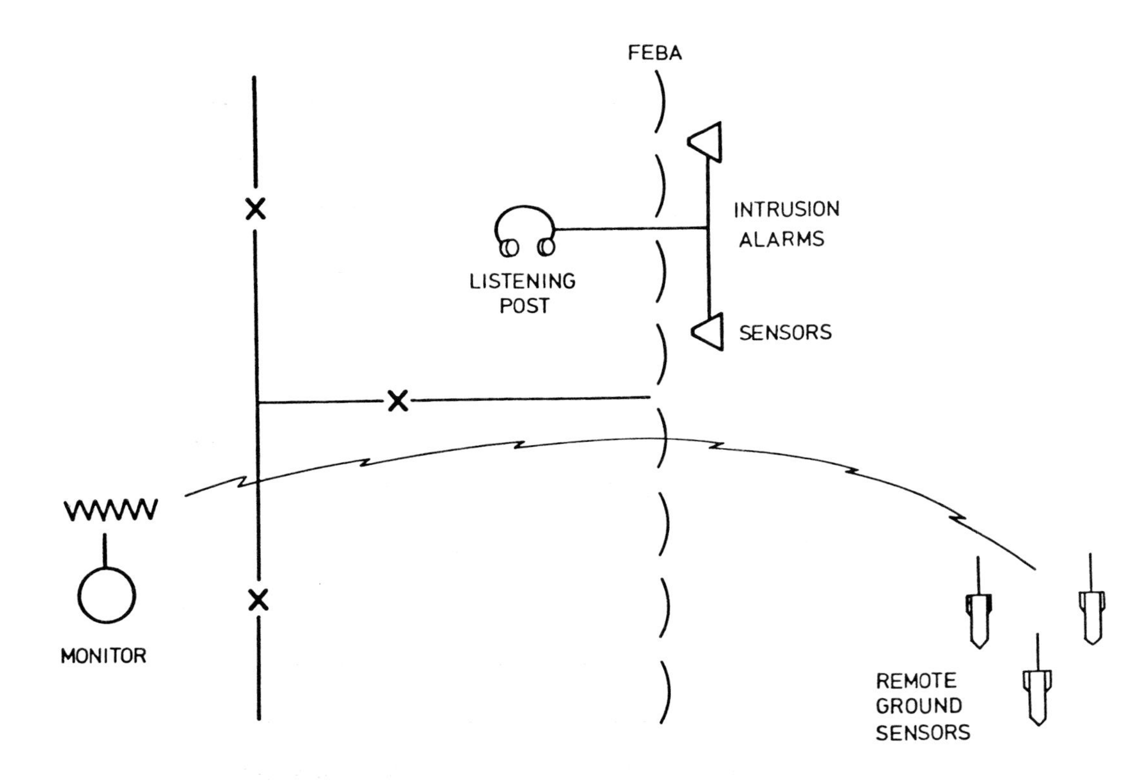

Fig. 7.8 Schematic layout of a RGS system

SUMMARY

There is a wide range of surveillance and target acquisition systems now available to give the modern battlefield commander a comprehensive capability for the detection, recognition and identification of targets in depth. With careful deployment of these resources he can maintain a watch on the enemy 24 hours a day and in almost any weather conditions in order to react early to the build up of forces against him. Recent advances in technology have improved the performance and reduced the size of these systems and it is likely that the next decade will see rapid advances in surveillance techniques, especially in the use of stand-off airborne systems and RPVs.

SELF TEST QUESTIONS

QUESTION 1 What is the average ground visibility in NW Europe?

Answer ....

QUESTION 2 Why are kites not satisfactory surveillance platforms?

Answer ....

....

....

QUESTION 3 What are the main characteristics of drones?

Answer ....

....

....

....

QUESTION 4 What are the essential features of RPVs?

Answer ....

....

....

....

QUESTION 5 What is the definition of Response Time of a drone?

Answer ....

....

QUESTION 6 List three types of sensors used in Remote Ground Sensors.

Answer ....

....

....

QUESTION 7 Why are satellites not widely used for tactical surveillance?

Answer ....

....

....

....

QUESTION 8 When was the use of aerial observation first recorded?

Answer ....

QUESTION 9 What are the two main factors which have led to the diversification of surveillance techniques from manned to unmanned aircraft?

Answer ....

....

QUESTION 10 What devices may be mounted on the tethered platform, such as KIEBITZ?

Answer ....

....

....

....

ANSWERS ON PAGE 195

8.

Counter Surveillance

GENERAL PRINCIPLES

Counter surveillance directed against enemy surveillance devices may have two objectives: negative counter surveillance aims to frustrate the attempts by the enemy to use his equipment to obtain information of the disposition, movement, and tactical activity of his opponents; positive counter surveillance seeks to inject false information into enemy surveillance devices so that he draws wrong conclusions concerning his opponent's activities and is led into tactical disaster.

In this chapter there will be much contrasting of the action and counter action of two opposing sides. It will clarify description to maintain Red land as using surveillance devices and Blue land who practises counter surveillance. This convention will be continued throughout the chapter.

If Blue land is to provide effective counter surveillance (CS) against Red land, his first requirement is for effective intelligence so that he knows the spectrum of devices being employed by Red land, and if possible the details of the equipment likely to be deployed against him. In the absence of such intelligence, Blue land must assume the enemy to have a capability at least comparable with his own.

Blue land must plan to spread his CS resources right across the range of Red land surveillance devices (SD) to avoid a dangerous gap. For example there would be little point in camouflaging a missile launcher position against visual observation if the thermal signature of the installation was allowed to remain prominent.

The sudden and obvious use by Blue land of CS should only be undertaken as a result of a deliberate tactical plan. Such an event would obviously lead Red land to expect some form of immediate tactical activity. This is of course merely an example of the universal concept that in warfare no single aspect of tactical activity must be allowed to be considered in isolation.

Another general consideration is the need to relate each CS technique to its immediate environment. This is very obvious when the impact of a jungle camouflage uniform in a snow covered land scape is considered, but is equally important when the seismic signature of mule transport against a background of modern motor transport is involved. CS also shares, with electronic warfare in general, the need for rapid and flexible response to unexpected situations created by Red land. The introduction of new equipment or radically new methods of using existing equipment are examples of this.

Finally the overriding importance of adequate training and self discipline on the part of Blue land troops cannot be overemphasised. The most excellent CS equipment will be largely ineffective if improperly used.

THE TECHNIQUES OF COUNTER SURVEILLANCE - THE VISUAL REGION

Most techniques of CS in the visual part of the spectrum are older than war itself. No discussion of CS would however be adequate without a short review of the main methods which have been developed over the years. These methods are also an important introduction to techniques used in other parts of the spectrum where the forms of visual CS have their counterparts; in many cases almost exact analogues. The point should be made however that a given technique may not have the same importance in different parts of the surveillance spectrum. For example Stand Off Jamming (SOJ) is highly effective in the disruption of microwave radar but is much less effective in the visual part of the spectrum.

The various visual CS techniques which Blue land may adopt in order to confuse or deceive the SD of Red land will now be discussed one by one.

The Adoption of Low Profile Movement and Deployment

This technique is nothing more nor less than the intelligent exploitation of the environment and in particular the use of ground by the Blue Commanders. The basic concept is the use of natural features to interrupt the line of sight between elements of the Blue land forces and probable points of location of the Red land SDs. Note must also be taken of the fact that moving objects are more readily detected than stationary ones and that regular or highly ordered movement is easier to locate than erratic movement. A dramatic example of the suppression of movement was the "freezing" of no mans land patrols when star shells were fired during the trench warfare of the first world war.

The well known scientific fact that regular patterns - in almost any context - are easier to recognise than irregular shapes or time profiles is now well established in military doctrine although history is rich in examples of the breach of this simple tactical concept.

In the context the CS task of avoiding obvious change in successive Red land photo reconnaissance pictures is an example of a particularly difficult CS problem.

The Use of Camouflage

There are two main aspects of the use of camouflage by Blue land. Firstly the task of covering elements of his forces to change their reflective characteristics so as to minimise the chance of detection by Red land SD. Secondly to devise methods and equipment to reduce the impact of Blue land activity on his environment of such a nature as to draw attention to that activity. The manufacture of "camouflage nets" which are effective in different situations is a highly skilled engineering technique capable of considerable further development, particularly when the need to camouflage at all frequencies of the electro magnetic spectrum is considered.

The ease with which the reflective covering can be deployed or packed for movement is of considerable importance. The crudity of currently available systems when compared with what nature has achieved in the chameleon and similar creatures is striking.

Damage to vegetation and ground by moving Blue land units, particularly heavy vehicles, is difficult to conceal. A more fruitful approach may well be to hide the movement of important military units among the normal flow of routine traffic. This is of course much easier to achieve on hard roads and in built up areas. A particularly interesting example of this type of CS is the proposal by the USA to conceal the whereabouts of its new MX missile by an entirely artificial environment which does not change when the missile is present.

The tendency of all human beings to walk the shortest distance possible is likely to produce highly characteristic weapon site signatures if track discipline is not firmly enforced.

THE USE OF BARRIERS AND DECOYS

The use of smoke screens to avoid visual observation has been used for centuries. In its modern form the rapid deployment of smoke by projectors is an important part of the self defence armoury of Blue land forces seeking to avoid observation by Red land SDs. It must be remembered however that although smoke may be highly effective against visual observation it is technically much more difficult to produce smoke barriers which are effective against Far IR systems or longer wavelength devices. This is due to the need to employ larger and larger suspended particles as the wavelength of concern is increased. Such large particles tend to sink through the air quickly making the production of persistent smoke clouds difficult. Considerable research is being mounted by various countries into this problem so that the situation could change in the next few years.

The use by Blue land of decoys includes the normal tactical ploy of "creating a diversion" which requires the use of normal forces and equipment to act as decoys. However, specially constructed equipment may also be used particularly to effect positive CS against Red land photo reconnaissance. Such devices may consist of inflatible models of the vehicles or equipment to be represented. Again it must be remembered that a dummy unit of this type may need to be provided with fake communication transmissions if the Red land force supports his photographic sensors: with electronic support measures (ESM). The decoys may also have to be provided with credible thermal emission characteristics. As with all other aspects of military hardware the impact of modern technology is such as to make everything much more costly.

Fig. 8.1 Armoured vehicle firing smoke projectors

THE PHYSICAL DESTRUCTION OF SURVEILLANCE DEVICES

Given the fact that Blue land knows the location of the Red land SDs, then the use of normal weapons to destroy those SD or the men who operate them is an obvious option for the Blue land forces. In the visual region the SD will in most cases be the human eye aided by various optical equipment. In the majority of circumstances the Red land visual SDs will be passive (ie they will not need to illuminate their targets) and their location may be in doubt. Where large SD installations are involved, Blue land may resort to raiding parties to destroy such installations in situ. In such a case expert advice and training should be available to ensure that permanent damage may be inflicted as economically as possible and that the same damage is inflicted on like installations so that subsequent "cannibalisation" by Red land may be frustrated.

In the tradition of the tactic of throwing sand into the eyes of the enemy, there is a current interest in the development of weapons specifically designed to destroy optical surveillance devices.

DAMAGE LASER WEAPONS

There has been much discussion in the national press of laser weapons of such power that they are capable of causing physical damage to strategic missile and other types of weapons. These are by their very nature subject to understandable security classification and it would be pointless to discuss such matters further.

However, it may readily be derived from information freely available in any physic's laboratory that if a visual or IR system is able to operate by reflected or radiated energy from a target at range R, then a laser, located at or near that target and operating at the same optical wavelength, would be able to direct energy of such a level as to destroy the detector of a system; whether the human eye or a semi-conductor device.

Surveillance system protective devices intended to minimise the effects of damage lasers would obviously be called for in the specification of any optical or IR device used for surveillance. The performance of damage lasers or counter counter devices would be closely guarded secrets.

THE USE OF NUCLEAR DEVICES

If the type of warfare being waged by Blue land and Red land includes the use of nuclear weapons the vulnerability of the Red land SD to nuclear attack must be considered. The normal effects of blast, heatflash and radiation will affect the Red land surveillance devices as it will affect any other equipment. In addition to this, Red land surveillance devices are likely to contain electronic signal processing circuits which will be particularly susceptible to any large electro magnetic fields or nuclear particle flux which may be created by Blue land, or indeed Red land, nuclear explosions. The hardening of electronic equipment against this type of attack is a highly specialised subject which will not be pursued here.

RETROREFLECTIVE PROTECTIVE BARRIERS AND DAZZLE LAMPS

There is plenty of historical evidence that the tactical use of the sun as a means of dazzling the enemy has been employed for a long time. There have been accounts of armies who have used highly polished shields to reflect the image of the sun into the faces of advancing enemy forces.

The extension of such concepts by Blue land to produce artificial lights or retro-reflective shields to throw back laser light must certainly be considered. However there is little published data on the current use of such systems which must therefore be considered speculative at the present time.

Where Red land employs image intensifiers however it is probable that Blue land would employ powerful light emitting flares which would cause such systems to overload and thus fail to function properly.

SUMMARY OF OPTICAL CS EQUIPMENT

To summarise the above discussion it may be said that in spite of the various counter optical techniques that have evolved over centuries of warfare, modern armies have little CS equipment for use in the visual region in general use. Their normal equipment is likely to be confined to camouflage and smoke producing equipment. It remains to be seen whether the damage laser will produce a significant change in the future.

COUNTER SURVEILLANCE TECHNIQUES AGAINST COMMUNICATIONS ESM

The extensive use of electro magnetic telecommunications by all modern armies leads to massive effort being put into surveillance devices which are effectively ESM receiving systems directed against telecommunication transmitters. Thus Red land would deploy ESM facilities which would not only monitor messages transmitted by Blue land, with the hope of decoding them and obtaining their actual meaning, but also would determine the signature and location of each transmitter with the intent of discerning some meaningful tactical pattern. Blue land would assume that such activity was in progress and might well go to considerable lengths to inject false information into the Red land system. This action and counter action is the essence of communications electronic warfare and the reader is directed to the telecommunications volume for further discussion of this vitally important aspect of modern war.

TECHNIQUES OF CS AGAINST MEDIUM LEVEL SURFACE TO AIR RADAR

The Problem

Medium level surface to air radars in the land force context are usually associated with anti-aircraft weapon systems. We are not here concerned with the counter systems operating against the weapon systems themselves but only those directed against the surveillance elements of them. Such systems currently deployed usually use missiles rather than guns and are typified by the American Improved Hawk equipment or the Russian SA6 mobile units.

For the purposes of this discussion we must consider the use by Blue land of CS to protect its aircraft against observation by Red lands Medium Anti-Aircraft Radar (MAR) which corresponds in role (but not in its hypothetical performance) to the improved Hawk Pulse Acquisition Radar (PAR) or the Russian "Flatface" Radar.

It is convenient to make some hypothetical assumptions about the performance of the Red lands MAR representing existing in service technology. It is assumed to be a low p.r.f. single beam radar with a good detection performance against small aircraft out to 100 Km range and 30 Km height. The beam sweeps a complete azimuth circle in 8 seconds.

Moving target indication (MTI) circuits are available to the operators who also have at their disposal certain anti-jamming facilities - unknown in detail to Blue land at the outbreak of hostilities. It is further assumed that information is obtained from the radar by operators observing targets on Plan Position Indicator (PPI) displays. The pulse length of the radar is 4 μsec giving a range cell of about 600 metres length at all ranges. A half power beamwidth in azimuth of 4° (70 mils) gives a resolution width in azimuth of 7000 m at the maximum range of 100 Km. In order to avoid ambiguous range problems the pulse repetition frequency of the radar may be taken as 1000 pulses per second or slower.

Faced with a number of Red land surface to air weapon systems equipped with such a radar, Blue land may adopt several counter surveillance activities. The first and most obvious technique is for Blue land aircraft to fly fast and low thus denying the Red land radar clear sight lines and hiding the aircraft echoes in ground clutter. The overall effectiveness of this tactic will, of course, depend on the potency of the Red land low level air defence.

Destruction of the Medium Anti-Aircraft Radar

The powerful transmissions of the MAR will enable Blue land ESM receivers to locate their position and the accuracy of the location can, if necessary, be enhanced by the use of additional sensors. Airborne ESM equipment will be the most effective and flexible but ground based equipment may also be used - particularly if high ground overlooking the MAR positions is available. The signature of the MAR will be highly characteristic and the Blue land ESM installations should have little difficulty in identifying them. Conditions on the battlefield may result in a very high density of pulses of electro magnetic energy so that modern ESM systems may well be capable of intercepting and processing up to a million pulses per second.

The Blue land task of location will be made more difficult if the Red land MAR has facilities for frequency agility, variable p.r.f. and adaptable pulse length. Once the position of the Red land MAR has been pin-pointed it may be attacked by any of the usual weapons including anti-radiation passive homing missiles which may be either air launched or surface launched.

Passive Jamming And Decoys

The use of passive jamming has been employed since the very early days of radar. The usual method is to distribute clouds of chaff, also known as "window", which consists of thin strips of aluminium cut to such a length as to produce a large reflection of the radar pulses. Blue land could "sow" chaff clouds from aircraft specially employed for the task or might use chaff shells or rockets.

The chaff clouds persist for quite long periods of time as the individual strips fall very slowly. The clouds move with the air that supports them and can cause general confusion to the Red land operators. However the use of MTI by Red land would greatly reduce the effectiveness of chaff. It will also be clear that the use of chaff in this type of engagement would require a carefully prepared Blue land plan. It would be of little or no value for the protection of an isolated aircraft against observation by a MAR.

Other forms of decoy involving the launch of relatively expensive airborne vehicles with specially enhanced radar echoing area may also be used in carefully pre-planned set piece attacks by Blue land, but it is unlikely that such a method would be used for routine tactical warfare or reconnaissance parties by individual aircraft.

Reduction of Radar Cross Section - The Stealth Approach

If the operating frequency of the Red land MAR can be assumed at the time of the development of the Blue land aircraft, much can be done to reduce the Radar Cross Section - or echoing area (RCS). The strength of the echo returned by a passive target such as an aircraft will depend upon its shape, the material of which it is made, and its aspect with respect to the MAR illuminating beam. In the past the design of the airframe and engines of aeroplanes has been dominated by aerodynamic considerations so that the reduction of RCS has been a matter of modifying the aircraft by minor shaping and the application of low reflective covering to selected areas at a later stage. However the vital importance of low RCS to the survival of the military aircraft is now fully appreciated so that consideration is given to this aspect of design right from the start, even though this must mean the reduction in performance in some other aspects of performance.

The development of radar absorbent material, and anti-reflectant paint has been in progress for many years and several commercial firms will supply them. However this is another example of an area where detailed information is difficult to obtain. This is also true of progress made in the performance of systems which seek to reduce target signatures by taking in the radar pulse by a special antenna, amplifying it, inverting its phase, and transmitting it back towards the radar so that the retransmitted pulse tends to cancel the pulse reflected from the skin of the target. This type of equipment, which may be regarded as a form of active jammer, could only be carried at the expense of other payloads.

Publicity given in the USA to the concept of the "stealth" type of strategic bomber tends to emphasise the importance attached to the reduction of the RCS of strategic and indeed tactical vehicles

The Use of Active Jammers to Defeat Medium Anti-Aircraft Radar

There are a number of different types of jammer which may be directed against a MAR and these must first be identified. The best and most effective jamming for Blue land to use will depend upon a number of factors which relate to the actual conflict under consideration. The first classification concerns the type of platform which carries the jammer. If the target itself carries the jammer it is described as a Target Borne Jammer (TBJ) or an Airborne Self Protection Jammer (ASPJ). If another Blue land aircraft - dedicated to jamming - provides jamming support for the target it will be called a Stand Off Jammer (SOJ) if it is further away from the victim MAR than the target it is trying to hide. If the jamming aircraft is close to the target it is called an escort Jammer (EJ) while a supporting jammer flying close to its victim is called a Stand In Jammer (SIJ).

The SIJ does not require very high jamming power because it is beaming its power into the Red land radar aerial at short range. It must however adopt a "loiter" flight path in the field of view of the victim radar. It is therefore likely to be a small cheap Remotely Piloted Vehicle (RPV) which may well be provided with a small warhead capable of destroying the radar as a "farewell gesture" when its fuel becomes exhausted.

The escort jamming aircraft, as it must remain close to the aircraft which it is protecting, is likely to be one with similar performance to the target. The SOJ on the other hand must stay on station behind the target aircraft for some time - probably in a racetrack shaped flight path. It is jamming from long range and must therefore provide considerable jamming power probably fed into large fairly narrow beam aerials. Such aircraft do not need high performance but they should have long endurance and large load carrying ability.

Any of the above jamming platforms may use either noise jamming or deception jamming.

Noise Jamming

Noise jamming is obtained by modulating a powerful transmitter, operating on the same frequency as the Red land MAR, with random signal fluctuation similar to those which cause random noise. In automatic radars this produces continual false alarms while display radars show a confusing PPI covered with fluctuating "speckles". If the Blue land jammers are fitted with ESM receivers the noise carriers can be accurately tuned to the exact frequency of the victim radars but if they are "blind" jammers the carrier of the jammer must be swept to and fro over the expected frequency band. Such swept jammers are less effective than the spot frequency type and partly for this reason many modern radars are frequency agile, that is the carrier frequency is changed in a random fashion from pulse to pulse. This reduces the efficiency even of those jammers provided with listening ESM receivers.

Deception Jamming

There are many types of deception jammer but the majority of them consist of a system which receives the transmitted pulse of the victim radar with a suitable ESM receiver. The pulse is then amplified and transmitted at an artificially delayed time. This gives the Red land operator a false target on his display at a range or position which differs from that of the real target. In some cases many false returns may be transmitted by the jammer which fills the Red land display with false targets. Such a jammer is termed a reverberation jammer.

Tactical Location of Jammers

In most circumstances the most effective form of jamming against a Red land MAR type radar would be a "circus" of SOJs placed by Blue land behind the attacking aircraft which were being concealed from detection. Such jamming - probably powerful noise jamming - should be positioned on the line extending from the position of the MAR through the location of the Blue land attacking aircraft. Such a set piece attack by Blue land would require careful planning with previous reconnaissance of the location of the Red land radars and the carrier frequencies likely to be used.

TECHNIQUES OF CS AGAINST LOW LEVEL SURFACE TO AIR RADAR

Surface to air radars intended for the detection and location of low flying aircraft, which must be considered to include both cruise missiles and small unmanned aircraft, are usually directly associated with the weapon systems designed to attack such vehicles. The parameters of these Low Level Radars (LLR) are quite different from those of the MAR because they must be able to operate against fast low flying aircraft which, in a ground environment, may "unmask" at very short range and remain in view of the radar for a few seconds only. Also the radar is subject to extensive ground clutter. These circumstances demand a radar with a high pulse repetition frequency so that blind approach velocities may be avoided, a pulse doppler form of signal processing so that good performance in heavy clutter can be achieved, a high rotation speed for the antenna which provides a high data rate, and automatic alarm and weapon activation to avoid human slowness of reaction and tendency to fatigue. On the other hand long range of detection is not required; a maximum of 15 Km or less. Faced with such a surveillance device the Blue land force would provide CS equipment intended to protect its low flying aircraft from attack by the weapon systems directed by the Red land LLR. The geometrical configuration of this type of situation does not lend itself to the SOJ since the long range lines of sight are unlikely to exist. On the other hand the automatic detection and essentially rapid response of the LLR lays it open to saturation by false alarms created by noise jamming from short range. Such jamming will of course alert the radar that an attack is probable but the excessive false alarm rate will prevent reliable location of the attacking aircraft. The platforms available to Blue land from which this noise jamming may be directed include an ASPJ pod on the attacking aircraft, a presensor RPV which may be either ground launched or drop launched from the Blue land aircraft, or a short duration "dump jammer" delivered by surface to surface projectile or rocket.

Such elaborate CS activity on the part of Blue land may seem excessively costly but it must be remembered that ground attack aircraft at present under development may well cost the equivalent of $30 million each (1981). Extension of their useful life by protection plans for each sortie may therfore be worthwhile. Where Blue land is proposing to mount a "stream" attack by a number of aircraft,

Anti-Radiation Missiles (ARMs) which home onto Red lands LLR may well be launched by the leading planes.

THE FUTURE

Much of the counter surveillance field is shrouded in secrecy and it is not possible to discuss openly details of current or proposed techniques. However, from the earliest times man has tried to hide himself and his equipment or deceive his opponent about his exact location. This will continue and it may be expected that as each new surveillance device is introduced a counter will be developed in time.

SELF TEST QUESTIONS

QUESTION 1 Why must all parts of the electromagnetic spectrum be considered when camouflaging equipment?

Answer ....

....

....

QUESTION 2 Why is the intelligent use of ground so important in counter surveillance?

Answer ....

....

....

....

QUESTION 3 Why should regular methods of activity be avoided if detection is to be made more difficult?

Answer ....

....

....

QUESTION 4 What is track discipline?

Answer ....

....

....

QUESTION 5 Why is it difficult to produce good smoke screens against IR sensors?

Answer ....

....

....

....

Answers to Self Test Questions

CHAPTER 1

Page 14

QUESTION 1 Surveillance is the continuous (all weather, day and night) systematic watch over the battlefield to provide timely information for combat intelligence.

QUESTION 2 A commander's area of interest includes the area occupied by enemy forces which could jeopordise the commander's mission. This is usually taken to be the area out to that occupied by the commander's opposite number's reserve. The Area of Influence is that area over which the commander can bring fire to bear.

QUESTION 3 6000 m.

QUESTION 4 An active system radiates energy in order to illuminate the target. It can therefore be detected and reveal the position of the observer.

QUESTION 5 The range of a surveillance system can be increased either by elevating the sensor (use of aerial platforms and high ground) or by moving the sensor closer to the target (patrols or RGS).

QUESTION 6 Radar is currently essential to a surveillance system because it is the only system with a truly all weather performance.

QUESTION 7 A complementary range of sensors is required to give a 24 hour, all weather system but it can only detect moving targets and its recognition ability is limited. An imaging system with higher resolution and good recognition capability is also required.

QUESTION 8 Contrast between a target and its surroundings is important because if there were no contrast, detection would be impossible. Reducing contrast is the aim of camouflage.

QUESTION 9 Seismic waves.
Acoustic waves.

QUESTION 10 In indirect fire systems the problem of location is more difficult because of the absence of a line of sight from the weapon to the target. The observer can locate the target, relative to his own position, using the same sort of techniques as for direct fire weapons. To transfer this information to the weapon it is necessary to know its position and that of the observer to the required degree of accuracy.

CHAPTER 2

Page 46

QUESTION 1
(i) Diffraction at the pupil.
(ii) Aberrations in the eye lens.
(iii) Finite size of rod and cone groups.
(iv) Defocusing.
(v) Ocular tremor.
(vi) Eye-brain interconnection and interpretation.

QUESTION 2 The vertical dimension of the screen is 11.4 inches and the number of line pairs per inch is $\frac{625}{2} \times \frac{1}{11.4} = 27.4$. Thus one cycle occupies $1/27.4 = 0.036$ inches and the range at which this subtends the limiting cycle of 1/60 degree of $\frac{1}{3438}$ radian is $0.036 \times 3438 =$ 125.5 inches.

QUESTION 3 Two line pairs is equivalent to 2 cycles and thus one cycle occupies one metre. Since the eye can resolve to 60 cycles per degree or to one cycle per 1/60 degree, ie one cycle per $\frac{1}{3438}$ rad the required range is that at which 1 metre subtends $\frac{1}{3438}$ rad viz 3.438 km.

QUESTION 4 a. At $3\ 10^{3}$ candela m^{-2} the Visual Activity is 0.9 cycles min^{-1}. The vertical angle subtended by the tank at 2km = 1 mrad = 3.43 min. The number of cycles subtended would be $0.9 \times 3.43 = 3.09$. According to the criteria the tank would be recognisable as such on a clear day. The visual angle would be $0.5 \times 1/0.9 = 0.16$ mrad and the minimum resolvable distance on the tank would be 0.32 m. Thus the turret and hull shape apertures should be discernable.

b. At $3\ 10^{-2}$ candela m^{-2} the visual acuity of the naked eye is 0.1 cycles min^{-1} and is improved to 0.7 cycles min^{-1} through a x 7 binocular. The number of cycles subtended by the tank in the vertical direction would be $0.7 \times 3.43 = 2.4$ cycles and the tank would still be recognisable as such on a clear day according to the criteria. However account would have to be taken of lens aberrations, loss of light through the binocular and movement of the binoculars.

QUESTION 5 The illumination of the film is proportional to the amount of light entering the camera which is proportional to the aperture area and hence the inverse square of the F-number. For a given blackening of film the product of illumination and exposure time is constant, and the exposure time must therefore be proportional to the square of the lens diameter. In the example quoted this virtually doubles between one F-number and the next higher one.

Other effects are larger depth of field and worse resolution at the higher F-numbers, assuming of course that the lens system has been fully corrected against aberrations down to the smallest F-number used.

QUESTION 6 In this problem the quoted eye resolution is 1 cycle in 0.5 mils or 2 cycles/mil.

Tank is 2.5 m high at 5 km and subtends 0.5 mils. For detection 2 cycles are needed hence tank needs to subtend 1 mil (eye resolution = 2 cycles/mil). As it only subtends 0.5 mils a magnification of $\frac{1}{0.5}$ = x 2 is needed.

Recognition needs 4 cycles, thus a magnification of twice the previous, x 4 is required.

For identification (8 cycles) a magnification of x 8 is necessary. 2, 4, and 8 cycles have been taken as necessary for detection recognition and identification rather than the slightly lower 1.4, 3 and 6.7 cycles more usually quoted. This apart from making the calculation easier allows for the inherent optical defects of telescope and binoculars even when they are firmly mounted. When, however, a binocular is hand-held, the efficiency is drastically reduced by hand tremor. Unfortunately hand tremor becomes more noticeable with increase in magnification.

Loss of telescope efficiency then requires higher magnifications for the task in hand, causing more hand tremor. Someone with fairly steady hands may well detect, recognise and identify the tank in this problem by hand holding binocular of x 2.5; x 5 and x 10-12. However in this sort of problem one assumes that the binocular is steadied if not firmly mounted in which case x 9 or x 10 may well give identification.

Note however that the ability to detect, recognise, and identify targets will be very much affected by target contrast.

QUESTION 7 As the existing telescope has a 100 mm focal length object glass, a replacement one of 200 mm f.1. would double the magnification. The eyepiece and rest of the optical system remain unchanged so as the magnification doubles the field of view halves to 4°.

In both cases the apparent field of view is the same, magnification multiplied by real field of view; 3 x 8° or 6 x 4° = 24°.

QUESTION 8 Originally the telescope was designed for use in poor light hence it would have an exit pupil of 6 or 7 mm diameter. The object glass would then have been, exit pupil x magnification, either (3 x 6) mm or (3 x 7) mm; (18 or 21) mm diameter. The replacement object glass would then have to be (6 x 6) or (6 x 7) mm =

36 or 42 mm diameter. Nowadays a 6 mm pupil 36 mm diameter object glass would be favoured.

For daylight use only, a 2.5 mm exit pupil could be used needing a 15 mm diameter object glass.

The F-number of the replacement object glass would be

$$\frac{\text{focal length mm}}{\text{diameter mm}},\ \frac{200}{36}\ \text{or}\ \frac{200}{15}, = 5.5 \text{ or } 13.3 \text{ respectively.}$$

QUESTION 9 The total subtense of the ship on the graticule is:-
40 + 30 mils = 70 mils.
Length of ship = 150 m.

$$\text{Distance of ship} = \frac{\text{length of ship x 1000}}{\text{number of mils}} = \frac{150 \times 1000}{70} = 2143 \text{ m.}$$

Length of image of ship on graticule is:-
focal length of object glass x number of mils ÷ 1000 = $\frac{150 \times 70}{1000}$
= 10.5 mm.

QUESTION 10 An exit pupil of 2.5 mm would be adequate for daylight use, hence

$$\text{magnification} = \frac{\text{object glass diameter}}{\text{exit pupil diameter}} = \frac{50}{2.5} = \times 20$$

The sort of eyepiece used in such a telescope would have an apparent field of view of about 40°.

$$\text{Hence real field of view} = \frac{\text{apparent field of view}}{\text{magnification}} = \frac{40^{\circ}}{20} = \frac{2^{\circ}}{}$$

A prismatic optical system similar to one half of a prismatic binocular, but with a longer focal length object glass would make for a compact telescope. The image brightness could be kept as high as possible by the use of efficient anti-reflection coatings on all lenses and prisms (blooming).

In time-to-come catadioptric optical systems (mirror lenses) may be used for small military observation telescopes.

CHAPTER 3

Page 66

QUESTION 1 To increase the illuminance level of the scene to a level at which the cones of the eye can operate and pick out enough detail to perform the required task.

QUESTION 2 About 10^5. This is the gain required to raise the illuminance level from overcast starlight to twilight when the cones of the eye begin to operate.

QUESTION 3 Moonlight, light reflected from the planets, starlight and sky glow.

QUESTION 4 To collect photons at visual and near infra-red wavelengths and convert them into electrons; to accelerate those electrons; to convert the higher energy electrons back into photons at visual wavelengths.

QUESTION 5 Photo emission occurs in certain substances when photons are absorbed and release electrons.

QUESTION 6 a. No emission occurs if the frequency of the radiation is below a certain threshold level whatever the intensity of the radiation.

b. The energy of the emitted electron depends only on the frequency of the incident photon.

c. The number of electrons emitted per second (the current) is proportional to the intensity of the incident radiation.

QUESTION 7 Advantages: Small size and weight.
High brightness.
Does not suffer from saturation and whiteout.

Disadvantages: More noisy at low light levels.
More expensive.

QUESTION 8 Low light television has the advantages of remote and multiple viewing and the ability to process the video signal. However, low light television systems will be heavier and more bulky than image intensifiers and consume more power. They are more expensive than image intensifiers.

QUESTION 9 There is a marked increase in the reflectivity of natural vegetation between 0.8 μm and 1.3 μm caused by chlorophyl. This means that vehicles painted in normal green paint may stand out against a woodland background. Infra-red reflectivity paint, the reflectivity of which increases at 0.8 μm, has been developed to ensure that camouflage is extended into the range of wavelengths covered by image intensifiers.

QUESTION 10 For short range work where size and cost are important constraints.

CHAPTER 4

Page 87

QUESTION 1 a. The hotter a body is, the more energy it radiates.

b. A hot body radiates over a greater range of wavelengths than a cooler body.

c. The peak radiation occurs at a shorter wavelength for a hot body than a cooler body.

QUESTION 2 The thermal radiation coming from any point in a scene is the sum of the blackbody radiation from the point plus the energy reflected by the point from neighbouring objects. Thermal contrast arises therefore, from variations in emissivity and reflectivity and these two tend to cancel out. However, since there is very rarely thermal equilibrium, contrast arises because of temperature variations in the scene.

QUESTION 3 Scattering and absorption.

QUESTION 4 No. Certain bands of frequencies are heavily attenuated. A surveillance system must use a working frequency outside the absorption bands and in one of the so-called atmospheric windows.

QUESTION 5 Because the energy of a far infra-red photon is less than the work function of any material known at present. Far infra-red detection utilises photo conduction and the pyro-electric effect.

QUESTION 6 Noise is the limiting factor. The noise must be less than the weakest signal it is desired to detect. Thermal noise and variations in the number of current carriers can be reduced by cooking. Cooling to cryogenic temperatures can reduce their effect to the extent that the device is limited only by the noise caused by the random arrival of photons.

QUESTION 7 An ideal detector would be made up of a mosaic of elements rather like the retina of the eye. This would have the advantage that no scanning would be required. However technology is not sufficiently advanced to allow the manufacture of half a million detector elements each with its own amplifier having identical responses.

QUESTION 8 The performance of a pyro-electric vidicon is inferior to that of a cooled thermal imager. Data rates are lower and spatial resolution is limited. Thus a pyro-electric vidicon is unlikely to have a range greater than 1 km but for short range work its advantages are that it does not require cooling (and can thus be left unattended for long periods) and its greatly reduced cost.

QUESTION 9 Not fully. Thermal imaging is a passive day and night surveillance system but its performance will be adversely affected by rain, fog, mist and high humidity.

CHAPTER 5

Page 121

QUESTION 1 (i) It is produced by stimulated emission and by population inversion.

(ii) It is coherent.
(iii) It is virtually monochromatic.
(iv) It is highly directional.
(v) It has very high beam irradiance.

QUESTION 2 (i) Active medium.
(ii) Chemical or electrical energy source to produce population inversion in the active medium.
(iii) Cavity resonator (cf microwave generator).
(iv) Telescopic beam expander if convergent output is required.

QUESTION 3 (i) A Q-switch to produce sharp single pulses for the best range accuracy.
(ii) A repetitive pulse output for range updating.
(iii) A range-gate.
(iv) Return signal detection with range gate stop pulse.
(v) A selection facility to eliminate unwanted targets.

QUESTION 4 (i) Much narrower beam divergence.
(ii) Higher target irradiance and stronger return signal per unit of transmitter power output.
(iii) Less background clutter.
(iv) Less prone to detection and to countermeasures.

QUESTION 5 (i) Laser light is more sensitive to atmospheric effects and weather.
(ii) There is an eye hazard in some cases.
(iii) It is more difficult to produce a high degree of coherence.

QUESTION 6 50 m diameter, which is impracticable.

QUESTION 7 Using the equation on page 106 for a pulsed laser the NOHD is 6.28 km.

QUESTION 8 This effectively reduces the source energy by 90% and the NOHD is reduced to 1.9 km.

QUESTION 9 Using the equation on page 106 for a CW laser the NOHD is 0.125 km.

QUESTION 10 Carry out the range practice as prescribed on UK Defence Standard 05-40 Issue 2 1977 (and updated amendments), noting in particular that:-

(i) A back-stop must be provided for all targets to be ranged with adequate side buffer zones, which must not transgress outside the controlled range area. Note the extra difficulty of containing the ruby laser beam.

(ii) The hazard area must be clear of personnel, livestock and specularly reflecting objects, including aircraft in flight.

(iii) Even though approved goggles may be worn by military personnel deliberate direct viewing must be forbidden and the whole of the hazard area must be cleared before firing even though the NOHD for goggle wearers has been reduced.

CHAPTER 6

Page 153

QUESTION 1 (c); none of the others is true.

QUESTION 2 a. 341 μs. Working; total distance travelled = 2 x 25, 550 m. ∴ Time taken = 2 x 25, 550 / 3 x 10^8 = 3.41 x 10^{-4} s or 341 s.

b. 0.33 μs. Working; total extra distance travelled to and from T2 = 100 m, corresponding to an extra elapsed time of 100/3 x 10^8 = 0.33 μs; for the 2 echoes to be distinct the pulse duration should be equal to or less than this.

QUESTION 3 (c); the radar must not transmit another pulse until echoes from targets at the extreme operational range have had time to return. This time is 2 x 5 x 10^4/3 x 10^8 s; hence p.r.f. must be equal to or less than the reciprocal of this time, ie 3 kHz.

QUESTION 4 AZ, BY, CW, DX.

QUESTION 5 a. At normal incidence it acts as a mirror, returning all the intercepted transmitted signal to the receiver (it has the same echoing area as a metal sphere of radius 2 A/ λ).

b. Similar to that for a conducting sphere, but σ is much smaller than the cross-sectional area because the diameter << λ . Nevertheless the aggregate of raindrops in a rain storm can produce a large rain clutter echo.

c. If the rod is parallel to the electric vector (E) it produces a large echo; conversely if the rod lies perpendicular to E, the echo is small. When parallel to E the rod acts as a short-circuit to the incident wave, causing a strong reflection. When perpendicular it has little or no effect.

d. The incident wave glances off the sloping sides and is lost, the only echo is from scattering from the point, and from the edge of the base.

QUESTION 6 (a) True; G α A α (linear dimension)^2, doubling both dimensions increases A 4-fold.

(b) False; beamwidth α λ /linear dimension ∴ each beamwidth is halved.

QUESTION 2 Because nearly all CS devices rely on the existance of clear lines of sight between CS sensor and target.

QUESTION 3 Because regular patterns in time or space are much easier to pin point than random activity.

QUESTION 4 The rigid training of troops not to create typical site signatures by always taking the direct path between two points on the ground.

QUESTION 5 Because IR devices use long wavelengths (typically 8 to 14 micrometres) and to obscure these wavelengths require large molecules which are relatively heavy.

QUESTION 6 Fake communications will also be needed to give a realistic presence to the decoy.

QUESTION 7 The carrier frequency, the rotation rate, the PRF . and the pulse length would be the most important.

QUESTION 8 Clouds of chaff may be carried by rockets, shells, mortar bombs, special carriers dropped by parachute or it may simply be fed through a hopper from a moving aircraft.

QUESTION 9 The shape may be designed for minimum RCS. Reflections may be directed away from the expected location of radar receivers by sloping of surfaces and radar absorbent material may be spread over parts of the structure likely to produce large backscatter. The airframe material can also be chosen to reduce reflection.

QUESTION 10 In principle CS measures are the same as those against MAR or LLR. Slight changes are necessary to allow for different geometry and the interference of sight lines by high ground. Ground based jammers may be very vulnerable to physical attack.

Glossary

GLOSSARY OF TERMS AND ABBREVIATIONS

A

A
: Effective aperture area of antenna.

APC
: Armoured Personnel Carrier.

ARM
: Anti-Radiation Missiles. Guided missiles designed to home onto enemy sensors.

Apparent Field of View
: A limitation imposed by the eyepiece design of the actual field of good optical definition. For any telescope the apparent field of view is the real field of view x magnification.

Area of Influence
: A geographical area in which a commander is directly capable of influencing operations, by manoeuvre and by fire support systems normally under his command and control.

Area of Interest
: That area of concern to the commander including the area of influence, areas adjacent to it and extending into enemy territory to the objectives of current and planned operations. This area also includes areas occupied by enemy forces who could jeopardise the accomplishment of the mission.

ATC
: Air Traffic Control.

Atmospheric Window
: Portion of the electromagnetic spectrum where the atmosphere is relatively transmissive.

AWACS
: Airborne Warning and Control System.

B

B
Receiver bandwidth.

Band Gap
Difference in energy levels between valance and conduction bands.

Beam Expander
A combination of optical elements which will increase the diameter of a laser beam.

Binocular
Literally two oculars; instrument with two parallel complete optical systems, one for each eye. Such an instrument should have considerable stereoscopic effect.

Biocular
In such a system two eyes look through one or two eyepieces at the image formed by a simple optical system, thus it has no stereoscopic effect.

Blind Approach Velocities
Radars with high p.r.f. are likely to overlook targets approaching at certain speeds.

Blooming or Coating
Surface treatment of a lens or prism surface to reduce reflection losses. Normally a piece of glass reflects about 5% of the incident light at each air/glass or glass/air surface. This can be reduced to as little as 0.2% with specialised blooming or about 1% with more normal techniques.

C

c
Velocity of electromagnetic waves (3×10^8 m/s).

CCD
Charge Coupled Device.

Chaff or Window
Strips of metal cut to the resonant length of a radar wave, intended to produce confusing reflections.

CLGP
Cannon Launched Guided Projectile.

Collimated Beam
: Effectively a parallel beam of light with very low divergence or convergence. A laser beam can be collimated much more effectively than other light beams since it is a coherent source.

Collimation
: The alignment of the optical axis of a telescope with respect to its mechanical axis. In the case of a binocular instrument the alignment of both optical axes with respect to an axis through the central hinge. Collimation error in the latter case shows, if large, as a double image or with small errors as eyestrain.

Collimation Sight (sometimes called lensatic sight)
: A relatively old idea, whereby a light spot is set at the focal point of a lens in a tube. This happens when one looks into the lens with one eye and the spot is seen apparently at a great distance superimposed on the view as seen by the other eye. There are many variations on the layout of this type of sight.

CMT
: Cadmium Mercury Telluride.

Conversion Index
: The gain of an image intensifier.

CRO
: Cathode Ray Oscilloscope.

CRT
: Cathode Ray Tube.

CW
: Continuous Wave.

CW Laser
: In which the radiation is sustained for periods greater than 0.1 second.

D

D
: Antenna width or diameter.

D*
: Measure of detectivity of a photon detector.

Damage Laser Weapons
: Lasers of such intense output power that they are able to destroy equipment illuminated by them.

Detection
The discovery by any means of the presence of a person, object or phenomenon of military significance.

Diffraction Limit
It is set by the degree of coherence which is attained. The smallest focused spot that can be obtained is when the laser output is derived entirely from the uniphase mode of oscillation.

Drone
An air vehicle which is normally unmanned and remotely or automatically controlled.

Dunp Jammer
An expendable jammer positioned by dropping from an aircraft or projecting from a gun or mortar.

E

E_i
Integration efficiency.

ECM
Electronic Counter Measures.

Entrance Pupil
Usually the object glass of a telescope or binocular but can be a real image of an internal stop formed by the object glass. In this case it can be located or focused in the same way as the exit pupil. By measuring the diameter of the entrance pupil and dividing by the exit pupil diameter the telescope's magnification is obtained. Sometimes a metal stop or aperture is placed immediately behind the object glass in poor quality telescopes; this then becomes the entrance pupil.

Erector
Lenses or prisms to erect normal inverted image formed by object glass.

Escort Jammer
A jammer carried in company with the target it is intended to protect.

ESM
Electronic Support Measure. Radio or radar receiving stations designed to analyse and locate enemy transmitters.

EW
Electronic Warfare. The action and counter action of two opposing sides to exploit the use of electromagnetic devices at the expense of the other side.

Eyepiece
The lenses by which the real image formed by the object glass is viewed.

Eye Relief (eye clearance)
Distance from last lens surface to exit pupil, where the eye has to be placed to see the full field of view. Long eye relief ('B' eyepiece) with fold-down eye cups aids spectacle wearers.

Exit Pupil
Image of the smallest angular stop as seen by the eyepiece. Once called the 'Ramsden disk' it can, except in Galilean telescopes, be located and focused onto a piece of white paper on translucent material just the eyeside of the eyepiece.

F

f
Frequency.

f_c
Carrier frequency.

f_D
Doppler frequency.

F
Receiver noise figure.

False Alarms
Spurious signals which falsely give the appearance of a target.

FEBA
Forward Edge of the Battle Area.

Frequency Agility
A technique for producing a different carrier frequency for each pulse of a radar, usually in a random fashion from pulse to pulse.

Field Stop
A metal ring usually at the focus of the eyepiece that forms a sharp boundary to the field of view.

Fixed Focus
Instrument set to an average focus, with preferably a sufficiently long eye relief to enable spectacles to be worn and correct the user's sight to normal.

FLOT
Forward Line of Own Troops.

FMCW

Frequency Modulated Continuous Wave.

G

G

Antenna gain.

Gas Laser

In this, laser action occurs by excitation, ionisation and molecular collisions in gas molecules.

Galilean

A telescope or binocular employing only a normal object glass and a concave eye lens. Such a telescope, although short, has a limited field of view for a given magnification. Galilean binoculars, once popular, are now only made as opera glasses.

Graticule or Reticule

Sighting or reference mark etched on glass plate and inserted into the field of view by placing it on an image plane. Then the graticule/ reticule is superimposed over the image. Originally spiders web crosses set in metal rings.

Ground Clutter

Confusion on a radar display caused by ground reflections.

H

Half Power Beam Width

The angular displacement used to describe the beam width of a radar.

I

Identification

The discrimination between object within a particular type or class, eg it is a General Abrams or T64 tank etc.

IFF

Identification, Friend or Foe. A system using radar transmissions to which equipment carried by friendly forces automatically responds thereby distinguishing themselves from enemy forces.

Improved Hawk

The medium level surface-to-air guided weapon currently deployed by NATO. A mobile system.

IR
Infra-red.

IRLS
Infra-red line scan.

IRR Paint
Infra-red reflecting paint.

IWS
Individual Weapon Sight.

K

k
Boltzmann's constant (1.38×10^{-23} J/K).

K
Kelvin degrees.

L

L
Loss factor.

Lambertian Surface
Diffuse surfaces, ie non specular or non mirror surfaces are usually considered Lambertian although they are seldom Lambertian at all angles. A Lambertian surface would have equal surface brightness at whatever angle from the normal it is viewed.

LED
Light Emitting Diode.

Lensatic
With lenses, referring to image erection by lenses or any form of sight that uses lenses. (See also collimation sight).

LEV
Laser Enhancing Viewing.

Light Transmission
The percentage of incident light emerging from an instrument's eye-piece.

LLTV
Low Light Level Television.

Location

A target is located when its position is fixed with the required accuracy for engagement.

Low Level Radars (LLR)

Radars designed to detect low flying aircraft.

Low Pulse Recurrence Frequency (Low p. r. f.)

A radar with long intervals between transmitted pulses such that no ambiguities in range measurement are likely to occur.

LRATGW

Long Range Anti-Tank Guided Weapon.

M

Magnification

The apparent linear or angular increase in image size. Sometimes superficial magnification on area magnification may be given. In this case 10 x linear or correct magnification become $10^2 = 100$ x area magnification.

MBT

Main Battle Tank.

Medium Anti-Aircraft Radar (MAR)

A hypothetical radar discussed in this book. Of a role corresponding to the acquisition radars of surface-to-air guided weapon systems.

Moving Target Indication (MTI)

A signal processing technique which reduces confusion by removing mobile targets from the radar display.

MX Missile

A long range strategic nuclear weapon under development in the USA.

N

N

Noise power.

Negative CS

Aims to prevent an opponent from getting information from his surveillance devices (SD).

NEP

Noise Equivalent Power.

NOD

Night Observation Device.

Nominal Ocular Hazard Distance

The distance beyond which the energy in the laser beam falls below the damage level for the eye.

O

Object Glass

Real image forming lens, front lens of normal telescope or binocular but set below top prism in periscopic type instrument.

Optically Pumped Laser

A class of laser which derives its energy from an incoherent light source such as a xenon flash lamp.

P

P

Transmitted power.

Photo Electron

An electron given off by a photo emitter when it absorbs a photon of radiation.

Plank's Constant

4.14×10^{-15} electron volt second.

Plan Position Indicator (PPI)

A form of radar display which shows targets at their measured bearing and slant range from the location of the radar.

Population Inversion

As the necessary condition for laser action, in which the population of the higher energy levels exceed that of the lower energy levels in contradistinction to that which occurs in nature.

Porro Prism

Normal, traditional type erecting prism as illustrated in Fig. 2.10.

Positive CS

Aims to inject false data into opposing sensors.

PPI

Plan Position Indicator.

p. r. f.

Pulse recurrence frequency.

Prism

Block of glass ground and polished accurately so that light is caused to deviate or be totally internally refracted.

Prismatic

Usually reference to an instrument having prism for image erection, ie prismatic binocular.

Protection Standard

This is the energy density (pulsed laser) or power density (cw laser) above which it is dangerous to the eye.

Pulsed Laser

This delivers its energy in the form of a single pulse or a train of pulses. When the p. r. f. is greater than 1Hz the laser is said to be repetitively pulsed.

Pulse Length

The time duration of the transmitted pulse of a radar - typically a few microseconds.

Q

Q-Switch

A device for producing very short intense pulses. Hence a Q-switched laser emits short, high-power pulses.

R

R

Range

Radar Cross Section (RCS)

The effective echoing area of a radar target usually expressed in square metres.

Range Finder

Instrument to determine range. Stereo, coincidence, stadiometric or unequal magnification (see text for description).

Real Field of View (Fov)

The actual angle of view as seen by the telescope or binocular.

Real Image

An image which can be formed onto a piece of paper or translucent material, eg ground glass.

Recognition

The determination by any means of the friendly or enemy character or of the individuality of another or of objects such as aircraft, ships or tanks, eg it is a tank/it is a lorry. It also refers to phenomena such as communications - or any electronic patterns

Reflex Camera/Sight

Camera in which the image is displayed on a focusing screen via a mirror. A reflex sight will also have a mirror perhaps only lightly silvered to reflect an aiming point into the field of view.

REMBASS

Remotely Monitored Battlefield Sensor System.

Remotely Piloted Vehicle (RPV)

An unmanned aircraft able to be controlled by a ground based or remote airborne pilot.

Resonant Cavity

The optical enclosure within which coherent energy is built up and from which emission results. The emission is pulsed or cw depending on the design and nature of the system.

RGS

Remote Ground Sensor.

RHI

Range Height Indicator.

Roof Prism

More complicated erector prism now used in modern slimline binocular. Smaller and usually more compact than porro prisms but have an edge on one face like a house roof that requires considerable manufacturing accuracy.

RPH

Remotely Piloted Helicopter.

S

S

Received signal power.

SAG

A Russian Surface-to-Air Guided Weapon roughly comparable with the improved Hawk.

Semiconductor Laser

In which laser action normally occurs in a junction diode. The output, which is relatively low, may be cw or pulsed.

Sensors. Surveillance Devices (SD)
Sensors or instruments used to obtain surveillance information.

SLAR
Sideways Looking Airborne Radar.

Solid State Laser
In which the laser action occurs in a glass or solid crystal, for example, Nd:YAG. Its operation relies on the presence of controlled impurity levels within the solid material. It is most commonly used as an optically pumped, pulsed laser.

SOTAS
Stand Off Target Acquisition System.

Stand Off Jamming (SOJ)
The use of confusing or deception jamming projected from a platform which is positioned behind the target being observed by a SD.

Stand In Jammer (SIJ)
A jammer placed much nearer to its victim than the target it is protecting.

Stealth Bomber
A bombing aircraft specifically designed to have a very low RCS.

Stereoscopic Acuity
A measure of the stereoscopic ability of the eyes.

Stereoscopic Effect
The three dimensional view of an image which is seen by two eyes rather than the flat view as seen by one eye.

T

t_o
Observation time.

T
Elapsed time.

T_o
Ambient temperature (taken as 290 K).

Target Borne Jammer (TBJ) or ASPJ
A jammer carried by the target it is protecting.

TGS
Tri glycine sulphate.

TI

Thermal Imager.

T/R

Transmit Receive Switch.

U

u

Target velocity.

V

Variable magnification or focal length

As zoom but the focus has to be adjusted each time the magnification or focal length is changed.

Veiling Glare

Diffuse light across the field of view giving an apparent reduction in image contrast.

Vernier Acuity

The ability of the eye to line up two vertical lines one above the other below a horizontal centre.

Virtual Image

An image which can be seen by the eye but not formed as a real image, ie the image seen through a telescope or microscope.

Visual Acuity

A measure of the resolving power of an eye in standard test conditions. This is taken as 1 minute of arc but varies greatly with light level.

Z

Zoom System

An optical system in which the magnification or focal length can be varied but the focus remains constant.

θ

Angle.

λ

Wavelength.

σ Radar echoing area.

τ Pulse duration.

ΔR Range discrimination.

Index

Absorption 72
Acoustic waves 7
Active systems 7, 69
AFV optics 40
Airborne surveillance 144
Aircraft,
  manned 158, 162, 164, 167, 169
  unmanned 162
Air defence 158-161
Alarm, 169
  false 169
Angle
  measurement 145
  resolution 147, 152
  tracking 145, 146
Antenna, 128, 134
  aperture area 136
  beamwidth 128
  cosecant^2 beam 131, 144
  fan beam 129
  faster scanner 135
  gain 136
  lightweight array 135
  multiple beams 133, 150
  phased array 149
  scan rate 129, 135, 148
Antimony 55
Anti-radiation missiles 182
Aperture stop 31, 32
Area of influence 1, 10
Area of interest 1, 10
Argon laser 99
ARGUS radar 159
Arms control 158
Artillery 161, 164, 168, 169
ATC 147
  codes 147
Atmospheric attenuation 138
  effects 95, 96, 97
  windows 72
Autokinetic illusion 27
Avionics 162, 168
AWACS 160

Balloons 158-159
Band gap 74
Bandwidth 80, 138
Battlefield surveillance 134
Beam divergence 94, 95
  focusing 95
Beta light 35
Binocular 37-39
Blackbody radiation 6, 69, 71
Blind approach velocities 182
  speeds 143

Cadmium activated zinc sulphide 56
Caesiated gallium arsenide 55
Caesium 55
Camouflage net 175
Carbon dioxide 99, 108
Cascade tube 56, 57
Cavity dump switch 102
  resonator 93
Chaff or window 179
Channel tube 57
Charge coupled device 62
Chirp 151
Chlorophyll 65
CL-89 162
CL-227 166
CL-289 163
CLGP 164
Clutter 134, 141
  filter 141
CMT 75, 79

Coherence 93
Coincidence range finder 40
Collimator sight 37
Combat intelligence 1
Cone adaptation 20
Conical scan 146
Continuous beam 91
Contrast 9, 71
  enhancement 60
Conversion index 56
Cooling 83
Countermeasures 7
Critical fusion frequency 27
  wavelength 74
CRO, CRT 126
CW radar 141

D* 76
Damage laser weapons 177
Dark adaptation 20
Dead time 127
Depth of field 31
Detection 3
Detector
  arrays 80
  response 80, 83
Dial sights 36
Digital data 146
  encoder 146
Direct fire simulation 114
Direction finding 125, 128
Displays 161
Doppler effect 125, 134, 141, 151
  frequency 141, 142, 143
  radar 141
  CW 141
  pulse 142
Drone 12, 162, 163, 164
Ducts, ducting 141
Dump jammer 182

Earth, observation of 158
ECM 163, 164
Elastic waves 7
Electro optics 52
Electronic beam steering 135, 145, 149
  range marker 132
  support measures 178
  warfare 174
Elevation coverage 129, 130
  measurement 150
Emissivity 9, 71
Entrance pupil 31, 32
Escort jammer 181
Europe, North West 157
EW 159, 163, 164
Exit pupil 31, 32, 35, 38
Explosions, detection of 158
Eye 17, 18
Eye relief 31, 32, 39

False alarms 181
FEBA 159-161, 168
Field of view 38, 164
Field stop 32
Fleurus, battle of 158
Fluorescence 56
FMCW 127, 128
f-number 31
Fortifications 158
Fovea 19
Frequency
  agility 179
  bands 128, Annex A

Gallium arsenide 64
  laser 104
Garbling 147
Gas laser 99
Gating 62, 64
Giraffe, radar 159
Ground clutter 179

Half power beamwidth 179
Height-finder radar 133
Helicopters 160, 165, 169
Helium-neon 99
Heterodyne detection 100
Ho Chi Minh trail 168
Holmium laser 103
Horizon-radra 139
HQs,
  artillery 161
  brigade 161
  divisional 161

Identification 3
  friend or foe 4
  target 161, 164, 168
IFF 147
  codes 147
Image intensifier 53, 56, 168
Image orthicon 60
Influence, area of 157
Infra-red 6, 12, 53, 55, 56, 60, 69
  72, 158, 159
  lenses 73
  reflecting point 65
Integration efficiency factor 138
Internal security 7
Interrogator 147
Ionisation 57
IRLS 166

Joule-Thompson cooling effect 84

Kerr cell switching 102
KIEBITZ 159
Kites 159

Lasers 7
  classification 105
  designator 164
  enhanced viewing 62
  range finding 105-108
  tracking 110-114
  weapons 114
Leiden-Frost 84
Lenses, photographic 158
Light adaptation 21
  units Appendix
Line of sight 4
Link,
  communication 162
  data 161, 163-166, 169
Liquid gas cooling 83
LLTV 159
Location 3
  target 164, 168
Losses, system 138
Low level radars 182
Low light television 60
Low pulse recurrence frequency 179
LRMTS 109
LTMR 109

Magnetism 128
Masts, folding 159
Medium Anti-aircraft Radar (MAR)
  178
Mesopic vision 22
Meteorological conditions 9, 65, 72
  85
Microprocessors 162, 169
MIDGE 162
Millimetre waves 152
  application 152
  attenuation 152
Missiles guidance 69, 78
  launching 158
Modulation transfer function 56
Monitors 168, 169
Mortar location radar 135
Moving target indication 179
MTI 143, 144, 166
Multiple pulse 91
  readout 60
MX missile 175

Navigation 164
  systems 148
Negative CS 173
Neodymium 64
  laser 101, 107
Night sky radiation 52
Nimrod in AEW configuration 144,
  145
Noise 75, 138
  equivalent power 76
  figure 138

Observation
  aerial 158
  time 138
Obstacles, man-made 157
Ocular hazard distance 104
Oculogravic illusion 27
Oculogyral illusion 27
Open night sight 35
Open sight 35
Optical range finder 40
Orbit, satellite 159
Oxygen 55

Passive radar 152

Passive systems 7, 69
Patriot 149
Patrols 168
Phosphor 56
Photo
  cathode 54, 56
  conduction 74
  emission 54, 60
Photographs, interpretation 162
Photon 92
  detector 74, 79
  performance 76, 80
Photopic
  response 51
  vision 19
Plan position indicator 179
Planck's Law 69
Platforms, tethered 159
Polarisation 82
  plane of 137
Population inversion 91, 93
Porro prism 102
Position CS 173
Potassium 55
PPI 147
  synthetic 133
p.r.f. 127, 143
  multiple 143
Projectile detection radar 135
Protection standard 103
Pulse
  comparison 151
  coded-phase 151
  Doppler effect in 151
  length 179
  linear FMCW 151
Pulse - Doppler 142
Pulse duration 151
Pumping cycle 98
Pupil reflex 23
Purkinje phenomenon 22
Pyroelectric vidicon 82

Q-switch 91, 102, 107, 111

Radar 7, 12, 159-161, 168
  basic characteristics 125
  cross section 180
  roles 125
  secondary 147
  3D 149
Range gate 64
Range measurement 125, 134, 146
Range resolution 127, 147, 151
Ranging 4
Rayleigh criterion 29
Reaction time 162
Receiver
  bandwidth 151
  noise bandwidth 151
  noise figure 151
Recognition 3
  target 161, 169, 170
Reconnaissance 158, 162
Reflection loss 140
Reflectivity 9, 64, 71
Reflectors 71
REMBASS 168, 169
Remote
  ground sensors 12, 78
  viewing 60
Remotely piloted vehicles 12, 181
Resolution 10
  criteria 34-
Responsor 147
Return signal 97
RGS 168-170
RHI 133
Rhodopsin 51
Rod adaptation 20
Rotors 159, 165, 166
RPH 165-167
RPV 162-169
Ruby 64
  laser 101, 107

SAG 178
Satellites 158, 159
Saturable dye switching 102
Saturation 58
Scanning 60, 62, 72
  parallel 80
  serial 82
Scattering 72
Scotopic vision 19
Search light 7
Secondary emission 57
  radar 147
Seismic waves 7, 12

Semi-conductor 55, 74
laser 103
Sensors
acoustic 169
electro-optical 159
elevated 159
infra-red 158, 159, 162, 168, 169
magnetic 169
optical 159
photographic 162
seismic 168, 169
TI 167
TV 164, 166
Ships 160
Silicon 61, 62
Silver 55
Single stage tube 56
Sky glow 52
SLAR 144
Sniffers 7
Sodium 55
Solid state laser 111
SOTAS 160, 164
Space 158
SPAHPLATTFORM 159
Spin-frequency jamming 146
Split beam 146
Spontaneous emission 92
Stadiometric range finding 40
Stand-off jamming 174
Stand-in jammer 181
Static split 146
Stealth bomber 180
Stephan-Boltzman's Law 69
Stephan-Boltzman's constant 70
Stereoscopic range finder 40
vision 26
Stimulated emission 91, 92
Surveillance 1
aerial 10
radar airborne 144
radar battlefield 134, 143, 150
remote ground based 10
radar surface-air 129, 143, 150
Surveying, secondary radar in 148
Survivability 161, 164

Target
acquisition 1
borne jammer 181
designator 108
echoing area 137, Annex B
emission 6
illumination by laser 110
TEA laser 100, 108
Terrain screening 10
Tethered platform 11, 12
Thermal
contrast 71
detector 73
imager 78, 168
pointing 76
radiation 52
spread 83
Threat 157
Tracking radar 145, 149, 150
Trajectories, orbital 159
Transmission 72
Transmitter 128
power 139, 151, 152
Transponder 147
Triglycine sulphate 82
Trilux L2A2 sight 36
T/R switch 132

U-2 aircraft 158

Vehicles 159
Vernier acuity of eye 26
Vidicon 60, 83
EB sicon 61
EB vidicon 61
pyroelectric 82
SEC vidicon 61
Vietnam 168
Visibility, limit of 157
Visual acuity 20, 23-25, 32
Visual search time 28

War 158, 159
Watergate 1
Weather 159
Wein's Displacement Law 69
Whiteout 58
Work function 55

Zone, artillery 164